KW-471-556

# Additional Scottish Executive Guidance on Aspects of EIA, June 2002

ENVIRONMENTAL IMPACT ASSESSMENT (EIA) DIRECTIVE

1) Minimum Requirements of the Regulations
2) Outline Planning Applications

ENVIRONMENTAL IMPACT ASSESSMENT: QUESTIONS & ANSWERS

# List of abbreviations

Some abbreviations are only used in Figures, Case Studies and Annexes to save space.

| | |
|---|---|
| AGLV | Area of Great Landscape Value |
| CCS | Countryside Commission for Scotland |
| DoE | Department of Environment (England) |
| EA | Environmental Assessment (the whole process) |
| EASR 88 | Former Environmental Assessment (Scotland) Regulations 1988 |
| EIASR 99 | Environmental Impact Assessment (Scotland) Regulations 1999 |
| EIA | Environmental Impact Assessment (the whole process) |
| EC | European Community |
| e.g. | For example |
| ES | Environmental Statement |
| IEMA | Institute of Environmental Management and Assessment |
| JNCC | Joint Nature Conservation Committee |
| LA | Local Authority |
| LNR | Local Nature Reserve |
| NCC | Nature Conservancy Council |
| NCCS | Nature Conservancy Council for Scotland |
| NNR | National Nature Reserve |
| NPPG | National Planning Policy Guidance |
| NSA | National Scenic Area |
| NVC | National Vegetation Classification |
| PAN | Planning Advice Note |
| PLI | Public Local Inquiry |
| RC | Regional Council |
| Reg. | Regulation |
| ROMP | Review of old mineral permissions |
| RSPB | Royal Society for the Protection of Birds |
| S. | Section (of Act of Parliament) |
| SAC | Special Area of Conservation |
| Sch. | Schedule |
| SCI | Site of Community Importance |
| SEA | Strategic Environmental Appraisal |
| SEPA | Scottish Environment Protection Agency |
| SINC | Site of Importance for Nature Conservation |
| SINS | Site of Importance to Natural Science |
| SMs | Scottish Ministers |
| SNH | Scottish Natural Heritage |
| SoS | Secretary of State |
| SPA | Special Protection Area |
| SPP | Scottish Planning Policy |
| SSSI | Site of Special Scientific Interest |
| TPO | Tree Preservation Order |
| UK | United Kingdom |
| WGS | Woodland Grant Scheme |

Scottish Natural Heritage

# A handbook on environmental impact assessment

**Guidance for Competent Authorities, Consultees and others involved in the Environmental Impact Assessment Process in Scotland**

Natural Heritage Management

Prepared for SNH by
David Tyldesley and Associates
Edinburgh

2nd Edition

2005

# Contents

## List of figures

# Part A

# Introduction to this Handbook

# A.1      Introduction to this handbook

★ Key information ★

## Box A.1.1

The Handbook is intended to provide Competent Authorities, statutory consultees and others involved in the Environmental Impact Assessment (EIA) process with practical guidance and a ready source of information about the process. In places it illustrates or concentrates on the treatment of natural heritage issues but, even where there is such a focus, the principles are often more widely applicable to other environmental topics. It is intended to help all of those involved in the process to make it more effective and therefore lead to better informed decisions.

**A.1.1**      This Handbook has been prepared and published in response to many of Scottish Natural Heritage's (SNH) partners expressing a need for a publication of this kind. It utilises the framework and content of an internal Handbook that was first prepared for SNH by David Tyldesley and Associates (DTA) in 1997, and which was extensively revised and reissued following the amendments to the legislation in 1999. This second edition of the Partners' version is based on a further major revision of the internal document, again by DTA. It also draws on the considerable experience that SNH has gained in participating in the EIA process.

**A.1.2**      References to important court cases and their implications have been added in the production of the second edition of this Handbook (see Annex 9). It has been prepared with all due care and diligence but it is not intended to be an authoritative interpretation of the law or government policy and neither DTA nor SNH can be responsible for any consequences from the use of the Handbook or any errors or omissions. Readers are advised to read the whole of the relevant court judgments and to seek their own legal advice in any particular case.

**A.1.3**      The Handbook is divided into six parts:

## Part A Introduction to this Handbook
This is a general introduction to the Handbook.

## Part B Introduction to the Environmental Impact Assessment Process
A general introduction to the EIA process, including the legislative background, the projects that are subject to EIA; and the contents of an Environmental Statement.

**A.1.4**      The rest of the main text of the Handbook then considers each of the main steps in the process of EIA, under four main stages: before the Environmental Statement is submitted; during the consideration of the Environmental Statement; the decision making stage and the post decision stage. Thus, the remaining four sections of the Handbook are as listed below.

## Part C Prior to the Submission of the Environmental Statement

Explaining the various stages before the Environmental Statement is submitted including deciding whether an Environmental Statement is required (the screening process); requiring the submission of an Environmental Statement; scoping an Environmental Statement; provision of information by consultees; baseline environmental information; predicting environmental impacts; assessing the significance of impacts; mitigating measures and enhancement; and presentation of environmental information.

## Part D Consideration of the Environmental Statement (and Project Consent Application)

Explaining the various stages of considering the Environmental Statement including consultation and publicity; liaison with the Competent Authority and the developer; wider consultation and dissemination; transboundary environmental effects; requiring more information or analysis; negotiating modifications of the project; Supplementary Environmental Statements; and reviewing the Environmental Statement.

## Part E The Decision Making Stage

Explaining the role of the Competent Authority and others and the stages of decision making, the roles of all the parties in these stages, including use of the precautionary principle; the relationship of EIA with the development plan and other consent procedures; and guaranteeing commitments and compliance with the decision of the Competent Authority.

## Part F Implementation and Compliance

Explaining the stages of implementation of the project and ensuring compliance with the terms of any authorisation given, in relation to mitigation and compensation for environmental effects, and the roles of the parties in these stages including time scale of implementation of mitigation and compensation measures; monitoring programmes; review, reassessment and remedial programmes.

**A.1.5**    There are then eight Annexes as follows:

Annexe 1    A Glossary of terms used in the Handbook
Annexe 2    List of Current Legislation, annotated
Annexe 3    List of Current National Policy and Guidance, annotated
Annexe 4    Projects Requiring Environmental Impact Assessment
Annexe 5    References and an Annotated Bibliography
Annexe 6    A brief résumé of the historical development of EIA in Scotland
Annexe 7    List of Principal Legal Cases Referred to.

**A.1.6**    The Handbook contains six Technical Appendices, which deal with detailed methodologies for impact assessment for:

Appendix 1  Landscape and Visual Impact
Appendix 2  Ecological Impact
Appendix 3  Earth Heritage Impact
Appendix 4  Impacts on Soils
Appendix 5  Outdoor Access Impact
Appendix 6  Effects on the Marine Environment.

**A.1.7**    Finally, the Handbook contains, in Attachment A, a 'master' copy of a scoping and review package to assist in scoping and reviewing Environmental Statements.

## Presentation

**A.1.8**    This Handbook covers a complex and often detailed range of information, policy advice, guidance and statutory and non statutory procedures relating to the whole of the EIA process. To make it more readable and easier to use, the text includes a series of figures and boxes. All of these are numbered for reference purposes.

The boxes used are as follows:

Blue-tinted boxes highlight or summarise key points of information.

★ **Key information** ★

Red-tinted boxes highlight key points of advice.

★ **Key advice** ★

Good EIA practice is highlighted in a green-tinted box.

★ **Good EIA practice** ★

## Application to Project Types

**A.1.9**    EIA is required for a wide range of project types. This Handbook applies to all project types in terms of the basic process of EIA. However, to continuously refer to all the different types of projects and different Competent Authorities and project proposers would make the text cumbersome and difficult to follow. For this reason and because the main body of the EIA guidance from government (Circular 15/1999) addresses the EIA process in relation to the town and country planning system, this Handbook tends to refer to 'developers' and Competent Authorities. Unless otherwise indicated the advice in this Handbook applies to the EIA process in respect of all project types, even though it concentrates on the main 1999 Regulations and the Circular relating to planning projects. Where a specific procedure relates only or primarily to planning authorities under the EIAR 99 then the term 'planning authority' is used instead of 'Competent Authority'.

## References to all Project Proposers as 'developers'

**A.1.10**    For the purposes of this Handbook, to help make the text more readable, all project proposers are referred to as 'developers', whether or not their project constitutes development within the meaning of the Town and Country Planning (Scotland) Act 1997 and whether or not the project is for public service or infrastructure or for commercial purposes.

## Scope of EIA Projects and Application of the Different EIA Regulations

**A.1.11**    Annexe 2 Table 1 of this Handbook lists all the relevant EIA Regulations relating to the different types of projects and their consent procedures. A summary of the scope of EIA regulations applicable to a wide variety of project types is given in Annexe 2 Table 2. This is followed by Table 3 which identifies the main project types in the various regulations and, for each one, summarises which is the competent authority; the relevant consent procedures; the relevant EIA Regulations; their geographical jurisdiction; the reference of the Statutory Instrument; and the date it came into force.

# Part B

## Introduction to the Environmental Impact Assessment Process

## B.1  Introduction to the EIA Process

★ Key information ★

**Box B.1.1**

**The EIA Process**

*'Environmental Impact Assessment (EIA)'* or, as the UK authorities used to refer to it, *'Environmental Assessment (EA)'* is the whole process of:

1. ● gathering environmental information;

2. ● describing a development or other project;

3. ● predicting and describing the environmental effects of the project;

4. ● defining ways of avoiding, reducing or compensating for the adverse effects;

5. ● publicising the project and the Environmental Statement including a clear, non-technical prediction of the likely effects, so that the public can play an effective part in the decision making process;

6 ● consulting specific bodies with responsibilities for the environment;

7. ● taking all of this information into account before deciding whether to allow the project to proceed; and

8. ● ensuring that the measures prescribed to avoid, reduce or compensate for environmental effects are implemented.

**B.1.1**    The **'Environmental Statement (ES)'** is the report normally produced by, or on behalf of, and at the expense of, the developer or project promoter which must be submitted with the application for whatever form of consent or other authorisation is required. It embraces the first four elements of:

1. gathering environmental information;
2. describing the project;
3. predicting and describing the environmental effects of the project; and
4. defining ways of avoiding, reducing or compensating for the adverse effects.

It is only one component, albeit a very important one, of the environmental information that must be taken into account by the decision maker. (See paras 8 and 21, Circular 15/1999 reference (17).)

**B.1.2**    The **'Environmental information'** that must be taken into account by the decision maker includes the Environmental Statement and all the comments and representations made by any organisation or member of the public as a result of the consultations and publicity that must be undertaken in every case. It also includes any further environmental or other information already held by the decision maker, which is relevant to the decision. (See para 8, Circular 15/1999 reference (17).)

**B.1.3**     Paragraph 6 of Circular 15/1999 (17) describes the EIA process as:

'The Directive's main aim is to ensure that the authority giving the primary consent (the 'competent authority') for a particular project makes its decision in the knowledge of any likely significant effects on the environment. The Directive therefore sets out a procedure that must be followed for certain types of project before they can be given 'development consent'. This procedure – known as Environmental Impact Assessment (EIA) – is a means of drawing together, in a systematic way, an assessment of a project's likely significant environmental effects. This helps to ensure that the importance of the predicted effects, and the scope for reducing them, are properly understood by the public and the relevant competent authority before it makes its decision.

★ **Key information** ★

**Box B.1.2**

It is important to appreciate that EIA is not, in itself, a decision making process.
It is a process that is integrated into existing decision making procedures, for example, the consideration of planning applications or woodland grant schemes, in order to better inform these decisions as to the environmental implications of the project. In this way, it contributes to the wider objectives of sustainable development.

**B.1.4**     Consequently, an EIA is never undertaken in isolation of some other procedure, indeed some procedures, such as the control of the intensive use of uncultivated land and semi-natural areas, were only introduced to provide a regulatory process to ensure compliance with the Directive. Comments made on EIA cases still need to focus strongly on representations as to whether the project should proceed, or how it should proceed.

★ **Key advice** ★

**Box B.1.3**

Comments on an Environmental Statement should be used to support and justify the representations made in respect of whether the project should be given consent, and if so, what conditions or limitations it should be subject to.

**B.1.5**     The advice in Box B.1.3 is fundamental to the process. This Handbook is designed to help contributions to the EIA process clearly distinguish between comments on whether the project should be consented and comments on the environmental information to be taken into account when the decision making body makes that decision. For example, it is perfectly possible that a consultee may find the conclusions of an Environmental Statement to be appropriate and acceptable but to conclude that the project ought not to be given consent. Equally, a perfectly acceptable project, from a consultee's point of view, could be accompanied by an inadequate and unacceptable Environmental Statement. In the latter case, the consultee would not, of course, object to the development, but may advise the competent authority about the inadequacy of the Environmental Statement.

**B.1.6** EIA is intended to ensure that the environmental effects of major developments and other projects likely to have significant environmental effects are fully investigated, understood and taken into account before decisions are made on whether the projects should proceed. Fundamental to the process are the statutory requirements for steps 5–8 in Box B.1.1 above, namely:

5. publicising the project and the Environmental Statement including a clear, non-technical prediction of the likely effects, so that the public can play an effective part in the decision making process;

6. consultation with specific bodies with responsibilities for the environment;

7. taking all of this information into account before deciding whether to allow the project to proceed; and

8. ensuring that the measures prescribed to avoid, reduce or compensate for environmental effects are implemented.

★ **Key information** ★

**Box B.1.14**

EIA should be of benefit to developers, decision makers and all of those consulted in the decision making process, including the public. It should help to ensure that development is sustainable, that development does not exceed the capacity of the environment to accommodate change without long-term harm. It should help to expedite the decision making process and guide the implementation of those projects that do proceed.

Many of the procedures are required by law but the effectiveness of EIA relies substantially on integrity and good practice.

**B.1.7** The process can be broken down into a series of stages and steps, which are reflected in the structure of the Handbook and summarised in Figure 1 below. Whilst the four main stages will normally follow consecutively, the steps within each stage could be undertaken concurrently or in a different order.

★ **Key information** ★

**Box B.1.5**

In practice, the whole EIA process should be an iterative one (repeated until the best solution has been found), with complex links back to earlier steps and a continuous process of assessment and reassessment until the best environmental fit is achieved.

**B.1.8** As PAN 58 (14) explains at paragraph 25:

*In practice the process rarely proceeds in a simple linear fashion. For example, environmental studies may identify a significant adverse impact which can only be overcome by altering the design, so the process reverts to the first step …*

**B.1.9** Not all of the steps in the process are actually required by law; some are a matter of good practice and common sense because without them the statutory requirements would be inadequate.

**B.1.10** It should also be noted that EIA procedures apply to projects in the marine environment; the procedures are not confined to land based developments in the way that statutory planning procedures are.

**B.1.11** The whole process is described in more detail in the following sections of the Handbook and the statutory and non-statutory elements are distinguished. The EIA process sits alongside decision making procedures and requirements. It does not directly duplicate other procedures, although it can be very closely related to them. For example, the decision making procedures required for a project that is likely to have a significant effect on a Natura 2000 site may use the information in an Environmental Statement, prepared under the EIA Regulations, in the appropriate assessment under Regulation 48 of the Habitats Regulations 1994. See Section E.2 of this Handbook and paras 80–81 Circular 15/1999 (17).

**B.1.12** Reference is made here to the various Annexes, Appendices and the Attachment at the end of this Handbook. To help illustrate and explain the EIA process, as it progresses through the Handbook, particular cross references are highlighted at the beginning of each Section.

| **Figure 1 Key Stages and Steps in the EIA Process** | |
|---|---|
| **Stage** | **Step** |
| **Stage 1:**<br><br>**Before Submission of the Environmental Statement** | Deciding whether EIA is required |
| | Requiring submission of an Environmental Statement |
| | Preliminary contacts and liaison |
| | Scoping the Environmental Statement |
| | Information collection |
| | Describing baseline environmental information |
| | Predicting environmental impacts |
| | Assessing the significance of impacts |
| | Mitigation measures and enhancement |
| | Presenting environmental information in the Environmental Statement |
| **Stage 2:**<br><br>**Submission of Environmental Statement and Consideration of Environmental Information** | Submission of Environmental Statement and project application for consent |
| | Consultation and publicity |
| | Requiring more information |
| | Negotiating modifications to the project |
| | Considering the environmental information |
| **Stage 3:**<br>**Making the Decision** | Making the decision |
| | Guaranteeing compliance |
| **Stage 4:**<br>**Implementation**<br>[For each of the pre-construction, construction, operational, decommissioning and restoration stages] | Implementation of mitigation and compensation measures |
| | Monitoring |
| | Review, reassessment and remedial measures |
| | Reporting |

## B.2   The Legislative Background

[See Annexes 2 and 4]

### The EIA Regulations

**B.2.1**      Annexe 2, Tables 1–3 of this Handbook list the current EIA legislation applicable in Scotland. It generally takes the form of 'Statutory Instruments' (Regulations), which are made by the Scottish Ministers, the UK Parliament or the Secretaries of State. Although not 'Acts of Parliament' they have much the same effect; they are statutory requirements. Failure to comply would render any case or decision open to challenge in the Court of Session, which means the decision could be quashed if it did not comply with the Regulations. References to important court cases and their implications have been added to the second edition of this Handbook (Annexe 9 below).

**B.2.2**      These statutory instruments are designed to bring into legal effect in Scotland the requirements of the EC Directives on EIA (4 and 19). These Directives have to be applied in Scottish domestic legislation in a way that is legally binding on developers and decision makers (decision makers are referred to in the Directive and the Regulations and in this Handbook as 'Competent Authorities').

**B.2.3**      The first Regulations appeared in 1988 with the *Environmental Impact Assessment (Scotland) Regulations* 1988 (5). These have been completely replaced by a new series of Regulations led by the *Environmental Impact Assessment (Scotland) Regulations* 1999 (18) (abbreviated in this Handbook to the EIASR 99) covering the majority of developments likely to require EIA on land in Scotland. These Regulations cover EIA requirements for:

● decisions on planning applications, appeals and deemed planning permissions made under the *Town and Country Planning (Scotland) Act* 1997 (6) (Part II of the Regulations);

● certain trunk road projects, comprising construction and improvement which are authorised under the *Roads (Scotland) Act* 1984 (12) (Part III of the Regulations);

● agricultural drainage works authorised by the Scottish Ministers by way of an improvement order under the *Land Drainage (Scotland) Act* 1958 (II) (Part IV of the Regulations).

**B.2.4**      Other Regulations cover a wide range of other project types and Annexe 2 Tables 1–3 below provide the full list.

**B.2.5**      As noted in paragraph B.1.10 above EIA procedures apply to projects in the marine environment. Consequently, there are important implications, for example, for marine fish farming, port and harbour developments, offshore dredging and wind farms and works requiring Marine Construction Licences under the *Food and Environment Protection Act* 1985 (37) (see Annexe 2 Tables 2 and 3).

**B.2.6**      The Directives (4 and 19) link EIA to the development consent procedure and therefore imply that all projects subject to EIA should require consent from a statutory authority before they can proceed. Since most of the project types listed in

Annexe I and Annexe II of the Directive already required some kind of consent under UK law the Government was generally able to implement the Directive by introducing sets of Regulations modifying existing legislation and procedures. Occasionally, however, it has been necessary to introduce new consenting procedures to meet the requirements of the Directives, for example, the control of the intensive use for agriculture of uncultivated land and semi-natural areas (see section B.5 below).

**B.2.7**    In addition to the suite of statutory Regulations, there are three other ways in which EIA, or an ES may be required:

1. The order making procedures under the provisions of S.14 of the *Transport and Works Act* 1992 (15) e.g. for major new infrastructure projects such as railways, tramways or bridges.

2. Parliamentary Standing Orders (Number 37A) (16) governing the procedures by which Private Bills for major development projects pass through the Scottish Parliament (see also Annexe 2 and Circular 26/1991).

3. By a Secretary of State or the Scottish Ministers introducing non-statutory guidance or procedures for development carried out by a Government Department or projects that may require the consent of a Secretary of State or the Scottish Ministers but which are not specified in the Regulations.

## The Power to Change the Regulations

**B.2.8**    Changes in EIA legislation relating to development are facilitated by Section 40 of the *Town and Country Planning (Scotland) Act* 1997 (6), empowering the Scottish Ministers to introduce further EIA Regulations, generally. The power includes the introduction of provisions different from the EIA Directives. As the Scottish Ministers could not make the Regulations less rigorous than the Directives, it follows that the power must have been introduced to enable a stricter regime than that directed by the EC, if the Scottish Ministers so wish.

**B.2.9**    The 1999 Regulations introduced a number of changes related to:

- widening the range of projects requiring to be assessed;
- taking account of Integrated Pollution Prevention and Control and integrating the provisions of the IPPC Directive 96/61/EC of 24.9.96 into the EIA process;
- the way in which potential international (transboundary) effects are to be considered;
- environmental interactions;
- screening the need for EIA and thresholds for determining whether assessment may be required;
- increased public information and accountability;
- scoping the content of the Environmental Statement;
- describing the alternatives considered; and
- applying assessment requirements to modifications and extensions of both Schedule 1 and Schedule 2 projects.

**B.2.10** In respect of the last bullet point, some Member States previously took the view that only modifications to projects in Annexe I were subject to EIA. However, the Courts have now held that modifications to Annexe II projects, as well as Annexe I projects, require EIA where they are likely to have significant environmental effects. European Court of Justice, ***Aannemersbedrijf PK Kraaijeveld BV v Gedeputeerde Staten van Zuid-Holland*** (October 24, 1996).

## B.3   The Projects that are Subject to EIA

[See Annexe 4]

### Statutory EIA: The General Principles

**B.3.1**      The Directive and Regulations relate to 'plans and projects' which require some form of licence, permission, consent or other authorisation before they can proceed.

**B.3.2**      Whether a project must be subject to the EIA process in Scotland depends entirely on whether it is of a kind listed in Schedules 1 or 2 of the Regulations issued by the Scottish Executive to ensure compliance with the EC Directives on EIA, as described in section B.2 above. The Schedules are reproduced in Annexe 4 below. Projects which are subject to the EIA process are of two kinds:

● those which are of a type of project that **must always** be subject to EIA, for example nuclear power stations (referred to as **Annexe I projects** because they are listed in Annexe I of the Directive, or now more widely referred to as Schedule 1 projects because they are listed in Schedule 1 of the Scottish Regulations) (see Annexe 4 of this Handbook); and

● those which may be subject to EIA if they are of a kind listed in Schedule 2 of the Scottish Regulations and Annexe II of the Directive, for example, a proposed urban development project.

If the development is of a type listed in Schedule 2 of the Regulations and meets one of the relevant criteria, or exceeds one of the relevant thresholds, listed in Schedule 2 of the Regulations, or is wholly or partly located in a sensitive area (see section B.4 below) it is referred to as **Schedule 2 development** and it must be screened to see whether it is, therefore, likely to have **significant effects** on the environment. If it is, it must be subject to EIA and is referred to as **EIA development** (see B.4, C.1 and Annexe 4 of this Handbook).

**B.3.3**      **EIA development** is development that must be subject to the EIA process because either it is a Schedule 1 project or it is a Schedule 2 project **likely to have significant effects on the environment** (for example because it meets one of the relevant criteria or exceeds one of the thresholds in Schedule 2 of the EIASR 99 or it is in a sensitive location and it is likely to have significant environmental effects).

**B.3.4**      The Scottish Executive Circular 15/1999 (17) provides advice in respect of determining whether a project of a kind listed in Schedule 2 is likely to have significant effects on the environment, including the publication of indicative thresholds for many of the Schedule 2 project types (see paras 17–18 and 28–47 and Annexe A Circular 15/1999 and Annexe 4 of this Handbook).

**B.3.5**      The Regulations do not bind the Crown but paragraph 165 of Circular 15/1999 explains that where a project that would otherwise require EIA is proposed by a Crown body, that body will submit an Environmental Statement with the consultation notice to the planning authority under the non-statutory consultation procedures set out in Circular 21/1984. The planning authority will then consider the proposal and the environmental information as if it was an ordinary planning application. See also B.3.10 below in respect of Private Parliamentary Bill procedures.

**B.3.6**    The Scottish Ministers have the power to direct that any particular project, or type of project, that would otherwise require EIA, is exempt from the requirement.

★ Key information ★

**Box B.3.1**

**Projects Requiring EIA**

All projects of the kinds listed in Annexe I of the Directive and Schedule 1 of the Regulations must be subject to EIA, in every case.

Projects of a kind listed in Annexe II of the Directive and Schedule 2 of the Regulations may need to be subject to EIA if the project exceeds certain criteria or thresholds or is in a sensitive area and is likely to have significant effects on the environment.

The Scottish Executive has provided guidance in Circulars and indicative thresholds for the nature, scale and location of particular Schedule 2 projects, which should be taken into account when assessing whether a particular project is likely to have significant effects.

However, whether or not a project exceeds the indicative thresholds, if it would not be likely to have significant effects on the environment it will not need to be subject to EIA.

But see further the commentary in section B.4 below.

### Voluntary EIA

**B.3.7**    The advantages of EIA are increasingly recognised by developers, some of which believe that an Environmental Statement can help to obtain a consent more quickly, especially where they consider the project to be environmentally benign.

**B.3.8**    An Environmental Statement may, therefore, be submitted voluntarily. That is, the project would not actually require to go through the EIA process, because it is not EIA development.

**B.3.9**    It should be noted that if an Environmental Statement is submitted to a planning authority (not to other competent authorities) – or a document referred to by an applicant as an Environmental Statement for the purposes of the EIA Regulations – the planning authority is required by Regulation 4(2)(a) of the EIASR 99 to treat it as an Environmental Statement and the proposal as EIA development even if it may not be (see further para 53 Circular 15/1999). Exceptionally, the planning authority may apply to the Scottish Ministers for a direction that the proposal is not EIA development if it is clearly not one to which the regulations apply, and processing the statement would be inappropriate (see last sentence para 53 Circular 15/1999).

### Parliamentary Private Bill Procedures

**B.3.10**    Article 1.5 of the Directive, and the Regulations, indicate that they do not apply to projects authorised or adopted by a specific Act of national legislation, such as Private Bills. There is a limited number of cases of these procedures in Scotland, mainly those relating to 'works' Private Bills for the Stirling–Alloa–Kincardine Railway, the Waverley Line and the two Edinburgh tram lines. In each case an Environmental Statement was prepared, but the full EIA procedure is not required.

## B.4 Criteria for Deciding whether EIA is Required
[See B.3, C.1 Annexe 4]

### Introduction

**B.4.1** Every Competent Authority has a duty to consider whether an application for any kind of consent that it receives for consideration is an application which should be subject to EIA (e.g. Regs 7 and 49 EIASR 99). If it is a Schedule 1 project EIA will always apply, unless it is 'exempt development' (see C.1.4 below for the definition of 'exempt development' and B.4.2 below for the definition of 'Schedule 1 development'). For other projects a two stage test is needed to determine whether EIA will apply.

a. Firstly, is the project a Schedule 2 project within the *set criteria* and thresholds in Schedule 2 of the Regulations and, if so,

b. Secondly, is it a Schedule 2 project likely to have significant effects on the environment by virtue of factors such as its nature, size or location (with *indicative thresholds* and criteria in Annexe A of the Circular)?

**B.4.2** 'Schedule 1 development' means *'development, other than exempt development, of a description mentioned in Schedule 1 of the Regulations'*;

'Schedule 2 development' means *'development, other than exempt development, of a description mentioned in Column 1 of the table in Schedule 2 where –*

*a) any part of that development is to be carried out in a sensitive area; or*

*b) any applicable threshold or criterion in the corresponding part of Column 2 of that table is respectively exceeded or met in relation to that development.'*

(Reg 2 EIASR 99)

**B.4.3** A 'sensitive area' is defined in Regulation 2(1) (see paragraph B.4.18 following below). It is stressed that development in a sensitive area should only be considered to be Schedule 2 development if it falls within a description in Schedule 2 (see Annexe 4 of this handbook).

### Determining whether an EIA is Necessary

**B.4.4** Generally, it will fall to competent authorities in the first instance to consider whether a proposed development requires EIA.

**B.4.5** Development outwith a sensitive area falling below the thresholds or meeting none of the criteria in the second column of the table in Schedule 2 does not normally require EIA and the authority need not adopt a screening opinion. In effect, the Regulations have already provided a negative screening opinion. However, there may be circumstances in which such small developments might give rise to significant environmental effects. In those exceptional cases Scottish Ministers can use their powers under regulation 4(8) of EIASR 99 to direct that EIA is required, even though it does not meet these thresholds and criteria. Such a direction will usually be in response to a request by the competent authority.

**B.4.6** It is emphasised that decisions need to be taken on a case-by-case basis. Thresholds shown within the indicative guidance in the Circular are not determinative. Individual projects that fall below the indicative thresholds and criteria in the Regulations may require EIA. The important thing is to consider whether the proposed development is likely to have significant environmental effects and to be clear about the reasons for the decision.

B.4.7 In legal proceedings, domestic courts must take account of judgements of the European Court of Justice (ECJ). So far as the EIA Directive is concerned the ECJ has consistently held that in its application it is to be interpreted as having a 'wide scope and broad purpose' **(Kraaijveld (Dutch Dykes) Case C-72/95)**. This has implications for Planning and other Competent Authorities when they are screening for EIA.

B.4.8 The wording of the EIA Directive should be interpreted widely. The fact that a particular type of development is not listed specifically within one of the categories of projects in the Directive or the EIA Regulations does not imply that it is not caught. The categories of projects are illustrative, not exhaustive. They should be read in a purposive manner to include similar types of project. Particular care is needed when considering development that could fall within the categories of 'industrial estate development' and 'urban development projects' listed under 'Infrastructure' projects (Schedule 2.10 projects).

B.4.9 A recent example of how the 'wide scope and broad purpose' applies is found in the Court of Appeal judgment relating to a planning proposal by the Big Yellow Property Company Ltd to construct a storage and distribution facility **(Goodman and another v Lewisham London Borough Council)**. The planning authority took the view that as such development was not specifically described in either the Directive or Regulations, there was no need to consider EIA. Following legal challenge, the Court of Appeal decided that:

*In this instance 'infrastructure' goes wider, indeed far wider, than the normal understanding, as quoted from the* Shorter English Dictionary, *of 'the installations and services (power stations, sewers, roads, housing etc.) regarded as the economic foundations of a country'.*

It held that the decision that the development was outside the reach of Schedule 2.10(b) of the EIA Regulations was outside the range of reasonableness that was open to the planning authority. The planning permission was quashed and the application remitted to the planning authority for reconsideration.

B.4.10   Thus, the Directive is **not** open to narrow interpretation. The UK Courts will interpret the Directive in the European sense – i.e. as having wide scope and broad purpose. It should not be assumed that a project is excluded simply because it is not expressly mentioned in either the Directive or the Regulations. For example, neither the Directive nor the EIA Regulations refer to specifically 'housing development'. But it would be a mistake to consider that housing development does not fall within the ambit of 'urban development projects'. Moreover, projects can be described in different ways so it is important to consider carefully the scope and purpose of the project—not just its label. A proposal to create a new 'Employment and Enterprise Opportunity Facility' may simply be another way of describing an industrial estate development.

## Changes or extensions to Schedule 1 or Schedule 2 developments

**B.4.11**   Changes or extensions to Schedule 1 or Schedule 2 developments also fall within the scope of the Regulations where the change or extension itself would fall within one of the descriptions in Schedule 1 or Schedule 2.

**B.4.12**   The criteria and thresholds in the second column of the table in Schedule 2 apply equally to changes or extensions to relevant development as they do to new development. Paragraph 13(a) of Schedule 2 provides that, in such cases, the thresholds and criteria are to be applied to the change or extension itself, not to the thing being changed or extended.

## The need for EIA for Schedule 2 development – general considerations

**B.4.13**   The Competent Authority must screen every application for Schedule 2 development in order to determine whether or not EIA is required. This determination is referred to as a 'screening opinion'. In each case, the basic question to be asked is: 'would this particular development be likely to have significant effects on the environment?' Section C.1 provides guidance on the screening process and related procedures. It should be read in conjunction with this section.

**B.4.14**   As a starting point, Schedule 3 EIASR 99 (see Annexe 4 to this Handbook) sets out the 'selection criteria' which must be taken into account in determining whether a development is likely to have significant effects on the environment. Not all of the criteria will be relevant in every case. It identifies three broad criteria which should be considered: the characteristics of the development (e.g. its size, use of natural resources, quantities of pollution and waste generated); the environmental sensitivity of the location; and the characteristics of the potential impact (e.g. its magnitude and duration).

**B.4.15**   In general, EIA will be needed for Schedule 2 developments in three main types of case:

a.  major developments which are of more than local importance;

b.  for developments which are proposed for particularly environmentally sensitive or vulnerable locations;

c.  for developments with unusually complex and potentially hazardous environmental effects.

**B.4.16**   The number of cases of such development will be a very small proportion of the total number of Schedule 2 developments. **It is emphasised that the basic test of the need for EIA in a particular case is the likelihood of significant effects on the environment.** It should not be assumed, for example, that conformity with a development plan rules out the need for EIA. Nor is the amount of opposition or controversy to which a development gives rise relevant to this determination, unless the substance of the objectors' arguments reveals that there are likely to be significant effects on the environment.

## Major development of more than local importance

**B.4.17**   In some cases, the scale of a development can be sufficient for it to have wide-ranging environmental effects that would justify EIA. There will be some overlap between the circumstances in which EIA is required because of the scale of the development proposed and those in which Scottish Ministers may wish to exercise their power to 'call in' an application for their own determination. However, there is no presumption that all called in applications require EIA, nor that all EIA applications will be called in.

## Development in environmentally sensitive locations

**B.4.18** The more environmentally sensitive the location, the more likely it is that the effects of a project will be significant and will require EIA. Certain designated sites are defined in regulation 2(1) as 'sensitive areas' and the thresholds/criteria in the second column of Schedule 2 do not apply there.

★ Key information ★

### Box B.4.1

All developments listed in Schedule 2 that may be located in the sensitive areas listed in regulation 2(1) and below must be screened for the need for EIA whether or not they meet the criteria or exceed the thresholds in Schedule 2. These are:

Sites of Special Scientific Interest
land to which Nature Conservation Orders apply
international conservation sites (e.g. SPAs, SACs and Ramsar Sites)
National Scenic Areas
Natural Heritage Areas (terminated by the *Nature Conservation (Scotland) Act 2004*)
World Heritage Sites
National Parks
scheduled monuments

(Note: Historic Gardens and Designed Landscapes are not listed and do not trigger the need for EIA)

**B.4.19** In certain cases other statutory and non-statutory designations which are not included in the definition of 'sensitive areas' may also be relevant in determining whether EIA is required. Circular 15/1999 at para 39 indicates that, where relevant, Local Biodiversity Action Plans will be of assistance in determining the sensitivity of a location. Urban locations may also be considered sensitive as a result of their heavier concentrations of population.

**B.4.20** Where statutory designations other than European or Ramsar sites are involved, including National Parks, SSSI, NNRs and NSAs, EIA will be appropriate where the particular natural heritage interest of the area would be likely to be significantly affected. Elsewhere, in the wider countryside it would be less likely that an Environmental Statement would be required on the grounds of the sensitivity of the location. However, the scale or nature of the proposal may be such as to require EIA, particularly if a major project is close to a human settlement.

**B.4.21** In considering the sensitivity of a particular location, regard should also be had to whether any national or internationally agreed environmental standards are already being approached or exceeded. Examples include air quality, drinking water and bathing water. Where there are local standards for other aspects of the environment, consideration should be given to whether the proposed development would affect these standards or levels.

★ Key information ★

**Box B.4.2**

**EIA Policy in Respect of International Designations**

Generally, Government policy, e.g. in NPPG 14, indicates that any Schedule 2 projects likely to significantly affect any of the following international designations (whether in them or not) will require an Environmental Statement to be submitted:

Classified and Potential Special Protection Areas;
Special Areas of Conservation/Sites of Community Importance; and
Ramsar Sites.

## Development with particularly complex and potentially hazardous effects

**B.4.22**  A small number of developments may be likely to have significant effects on the environment because of the particular nature of their impact. Consideration should be given to development which could have complex, long-term or irreversible impacts, and where expert and detailed analysis of those impacts would be desirable and would be relevant to the issue of whether or not the development should be allowed. Industrial development involving emissions which are potentially hazardous to humans or the natural environment may fall into this category.

## Indicative criteria and thresholds

**B.4.23**  Given the range of Schedule 2 development, and the importance of location in determining whether significant effects on the environment are likely, it is not possible to formulate criteria or thresholds which will provide a universal test of whether or not EIA is required. The question must be considered on a case-by-case basis. To assist in this, Annexe A of Circular 15/1999 sets out indicative thresholds and criteria. In the Scottish Ministers' view these offer a broad indication of the type or scale of development for which EIA is more likely to be required and, conversely, an indication of the sort of development for which EIA is unlikely to be necessary.

**B.4.24**  Annexe A of Circular 15/1999 also gives an indication of the types of impact that are most likely to be significant for particular types of development. It should not be presumed that developments falling below these thresholds could never give rise to significant effects, especially where the development is in an environmentally sensitive location. Equally, developments which exceed the thresholds will not in every case require assessment. The fundamental test to be applied in each case is whether that particular type of development and its specific impacts are likely, in that particular location, to result in significant effects on the environment. It follows that the thresholds should only be used in conjunction with the general guidance, and particularly that relating to environmentally sensitive locations.

## Applying the guidance to individual developments

**B.4.25** In judging whether the effects of a development are likely to be significant, Competent Authorities should always have regard to the possible cumulative effects with any existing or approved development. There are occasions where the existence of other development may be particularly relevant in determining whether significant effects are likely, or even where applications for development should be considered jointly to determine whether or not EIA is required.

*Multiple applications*
**B.4.26** For the purposes of determining whether EIA is required, a particular application should not be considered in isolation if, in reality, it is properly to be regarded as an integral part of an inevitably more substantial development. In such cases, the need for EIA (including the applicability of any indicative thresholds) must be considered in respect of the total development. This is not to say that all applications that form part of some wider scheme must be considered together. In this context, it will be important to establish whether each of the proposed developments could proceed independently and whether the aims of the Regulations and Directive are being frustrated by the submission of multiple or sub-divided applications.

★ **Key advice** ★

**Box B.4.3**

Competent authorities should press developers to submit complete projects and complete Environmental Statements to ensure that the aims of the Regulations and Directive are not being frustrated by the submission of separate applications, the key test being whether the proposed developments could proceed and fully operate as submitted.

*Changes or extensions to existing or approved development*
**B.4.27** Development which comprises a change or extension to Schedule 1 or 2 development requires EIA only if the change or extension is likely to have significant environmental effects. This should be considered in light of the general guidance in Circular 15/1999 and the indicative thresholds in Annexe A reproduced in Annexe 4 below, taken from Annexe A of Circular 15/1999. However, the significance of any effects must be considered in the context of the existing development. For example, even a small extension to an airport runway might have the effect of allowing larger aircraft to land, thus significantly increasing the level of noise and emissions. In some cases, repeated small extensions may be made to development. Quantified thresholds cannot easily deal with this kind of 'incremental' development. In such instances, it should be borne in mind that the criteria/thresholds in Annexe A of the Circular are only indicative. An expansion of the same size as a previous expansion will not automatically lead to the same determination on the need for EIA because the environment may have altered since the question was last addressed.

**B.4.28**   Competent Authorities are encouraged in the Circulars to consult other bodies, where relevant, when deciding whether the effects of a development proposal are likely to be significant and to take any views expressed into account.

## Outline planning applications (see further D.11 below)

**B.4.29**   Where it applies, the Directive requires EIA to be carried out prior to the grant of 'development consent'. Development consent is defined as 'the decision of the Competent Authority or Authorities which entitled the developer to proceed with the development'. Under the UK planning system, it is the planning permission that enables the applicant to proceed with the development. Therefore, where EIA is required for a planning application made in outline, the requirements of the Regulations must be fully met at the outline stage since reserved matters cannot be subject to EIA.

**B.4.30**   When any planning application is made in outline, the planning authority will need to satisfy themselves that they have sufficient information available on the environmental effects of the proposal to enable them to determine whether or not planning permission should be granted in principle. In cases where more information is required, authorities should request further information on the Environmental Statement under regulation 19 EIASR 99 and further information on the application, within one month of its submission, under article 4(3) of the *General Development Procedure Order* 1992 (6). Guidance on this stage is also provided in PAN 58 at paragraphs 28–31 and Circular 15/1999.

---

★ **Key advice** ★

**Box B.4.4**

It will be evident from the explanation in this section, and the collation of advice on outline planning applications in D.11 below, that all parties should ensure that all likely significant environmental effects are addressed at outline planning application stage and not left for approval of reserved matters.

---

**B.4.31**   The planning permission and the conditions attached to it must be designed to prevent the development from taking a form—and having effects—different from what was considered during EIA. This was confirmed in the case of *R V SSTLR ex parte Diane Barker* (2001).

**B.4.32**   The cases of *R v Rochdale MBC ex parte Tew* (1999) and *R v Rochdale MBC ex parte Milne* (2000) (Annexe 7) set out the approach that planning authorities need to take when considering EIA in the context of an application for outline planning permission if they are to comply with the Directive and the Regulations. Both cases dealt with a legal challenge to a decision of the authority to grant outline planning permission for a business park. In both cases an Environmental Statement was provided. In ex parte Tew the Court upheld a challenge to the decision and quashed the planning permission. In ex parte Milne, the Court rejected the challenge and upheld the authority's decision to grant planning permission.

B.4.33   In *ex parte Tew*, the authority authorised a scheme based on an illustrative masterplan showing how the development might be developed, but with all details left to reserved matters. The Environmental Statement assessed the likely environmental effects of the scheme by reference to the illustrative masterplan.

However, there was no requirement for the scheme to be developed in accordance with the masterplan and in fact a very different scheme could have been built, the environmental effects of which would not have been properly assessed. The Court held that the description of the scheme was not sufficient to enable the main effects of the scheme to be properly assessed, in breach of Schedule 4 of the Regulations.

**B.4.34** In *ex parte Milne*, the Environmental Statement was more detailed; a Schedule of Development set out the details of the buildings and likely environmental effects, and the masterplan was no longer merely illustrative. Conditions were attached to the permission *'to tie the outline permission for the business park to the documents which comprise the application'*. The outline permission was restricted so that the development that could take place would have to be within the parameters of the matters assessed in the Environmental Statement. Reserved matters would be restricted to matters that had previously been assessed in the Environmental Statement. Any application for approval of reserved matters that went beyond the parameters of the Environmental Statement would be unlawful, as the possible environmental effects would not have been assessed prior to approval.

**B.4.35** The judge emphasised that the Directive and Regulations required the permission to be granted in the full knowledge of the likely significant effects on the environment. This did not mean that developers would have no flexibility in developing a scheme. But such flexibility would have to be properly assessed and taken into account prior to granting outline planning permission.

**B.4.36** He also commented that the Environmental Statement need not contain information about every single environmental effect. The Directive refers only to those that are likely and significant. To ensure it complied with the Directive the authority would have to ensure that these were identified and assessed before it could grant planning permission.

**B.4.37** The Court of Appeal in *ex parte Diane Barker* (2001) confirmed this approach and there are some general conclusions that can be drawn about applications for outline planning permission:

a. An application for a 'bare' outline permission with all matters reserved for later approval is extremely unlikely to comply with the requirement of the Regulations.

b. When granting outline consent, the permission must be 'tied' to the environmental information provided in the Environmental Statement, and considered and assessed by the authority prior to approval. This can usually be done by conditions although it would also be possible to achieve this by a planning agreement (under section 75 of the Town and Country Planning (Scotland) Act 1997).

c. An example of a condition was referred to in *ex parte Milne* (2000). *'The development on this site shall be carried out in substantial accordance with the layout included within the Development Framework document submitted as part of the application and shown on (a) drawing entitled "Master Plan with Building Layouts'".'* The reason for this condition was given as *'The layout of the proposed Business Park is the subject of an Environmental Impact Assessment and any material alteration to the layout may have an impact which has not been assessed by that process'* (see paras 28 and 131 of the judgement).

d. Developers are not precluded from having a degree of flexibility in how a scheme may be developed. But each option will need to have been properly assessed and be within the remit of the outline permission.

e. Development carried out pursuant to a reserved matters consent granted for a matter that does not fall within the remit of the outline consent will be unlawful.

## The Degree of Confidence in Predicting Likely Significant Effects

**B.4.38**   The EIA Directive requires, amongst other things, firstly, that Competent Authorities decide whether EIA procedures apply to particular projects, a decision which in part is based on the likely significant effects on the environment; and secondly, that they take into account the effects before granting permission. At the first stage, the responsibility is to consider whether the project is likely to have a significant effect on the environment. This calls for the exercise of professional judgement taking into account factors such as nature, scale and location of the project (see Schedule 3 of the EIA Regulations), knowledge of the local area and its environment and evaluation of such information as it is reasonable to expect the applicant to provide at this stage. But the amount of information necessary at this stage does not mean you need to have 'full knowledge' of every environmental effect. Only if it is decided that EIA is required, will full and detailed knowledge of the project's likely significant effects be required.

**B.4.39**   A helpful judgement in this respect is that of ***Regina oao Jones v Mansfield DC*** where the judge held that in general a lesser degree of information is needed at the first stage of deciding whether EIA is required at all than at the second stage where it is necessary to provide the information. He commented that

*It is for the authority to judge whether a development would be likely to have significant effects. The authority must make an informed judgement, on the basis of the information available and to any gaps in that information and to any uncertainties that may exist, as to the likelihood of significant effects. The gaps and uncertainties may or may not make it impossible to reasonably conclude that there is no likelihood of significant environmental effects. Everything depends upon the circumstances of the individual case.*

**B.4.40**   The judgement also noted that

*Whether sufficient information is available to enable a judgement to be made as to the likelihood of significant environmental effects is a matter for the authority, subject to review by the court on Wednesbury principles.*

## Can the decision whether to require EIA take account of conditions and other measures that could ensure that likely significant effects were rendered unlikely or insignificant?

**B.4.41**   Conditions can still be used in granting permission to EIA development, but planning authorities need to exercise care and judgement to ensure that conditions designed to mitigate the likely effects of a proposed development are not used as a substitute for EIA or to circumvent the requirements of the EIA Directive. It may be useful to refer to relevant recent case law.

**B.4.42** *Regina oao Lebus v South Cambridgeshire DC* involved development for an egg production unit to house 12,000 free range chickens. A local resident had written to the planning authority in 2000 suggesting that EIA was required for this development. After a meeting and discussion with the applicant, the planning officers dealing with the case took the view that this was not EIA development and the applicant was told informally that EIA was not required. The planning officer dealing with the case made no written record of his conclusions. At the meeting the officers concluded that the potential adverse impacts of the development would be insignificant with proper conditions and management enforceable under a section 106 planning obligation (equivalent to a S.75 Planning Agreement in Scotland). Planning permission was granted subject to conditions in 2002. The resident challenged the decision by judicial review.

**B.4.43** The Court allowed the appeal and quashed the planning permission. So far as planning conditions and EIA are concerned it held

*it is not appropriate for a person charged with making a screening decision to start from the premise that although there may be significant impacts, these can be reduced to insignificance by the application of conditions of various kinds. The appropriate course in such a case is to require an environmental statement and the measures which it is said will reduce their significance.*

**B.4.44** The message from **Lebus** is that where proposed development is EIA development the use of conditions cannot be used to substitute for the proper assessment procedure. To do so would simply negate the purposes of the Directive. It is also clear from this case that planning authority staff need to make formal screening opinions on Schedule 2 applications.

**B.4.45** The question of planning conditions was also considered in *Gillespie v First Secretary of State and Bellway Urban Renewal*. In this case the First Secretary of State granted planning permission for a housing development on the site of a former gas works. One of the former gasholders was still in situ. Soil surveys on the site had been carried out and revealed contamination but the type and extent was not fully known, particularly of that below the gasholder. The First Secretary of State, however, considered that there was no need for an EIA. He permitted the development subject to conditions to carry out a detailed site examination to establish the nature, extent and degree of the site contamination and to remedy it prior to commencement of the development. The remediation strategy would rely upon tried and tested methods so there was no reason to assume they would be unsuccessful in removing the contamination.

**B.4.46** In quashing the First Secretary of State's decision, the Court of Appeal held that on considering whether an environmental impact assessment was required before planning permission could be granted the First Secretary of State did not have to ignore proposed remediation measures, but neither could he assume that, in a case of any complexity, they would be successfully implemented. The extent to which such measures could be taken into account in screening decisions would depend on the facts of each individual development having regard to:

a. the extent of the investigation into the impact of the development and environmental problems arising from it, up to the time of the screening decision;

b. the nature of the proposed remedial measures including uncertainties;

c. the extent to which those have been particularised;

d. their complexity;

e. the prospects of their successful implementation;

f. the prospect of adverse environmental effects in the course of the development, even if of a temporary nature;

g. the final effect of the development.

**B.4.47** *Gillespie* indicates that remediation measures need not be ignored when making decisions about the likely significant effects of proposed development. But care and judgement have to be exercised. Remedial measures that are well-established and uncontroversial, e.g. cleaning wheels of lorries and covering their loads to minimise dust etc., may well be taken into account. In more complex development, and/or where the nature of the proposed remediation measures is likely to be more complex and possibly less clearly established, it may be less appropriate to take the proposed measures into account. It is important that the offer of remediation measures is not used to frustrate the purpose of the EIA directive or serve as a surrogate for it.

**B.4.48** See also the cases of *R v Rochdale MBC ex parte Tew* and *R v Rochdale MBC ex parte Milne* in respect of outline planning applications at B.4.32 above.

## B.5　Provisions Introduced for Projects that Require a New Consenting Procedure

### Part A
### Use of Uncultivated Land and Semi-Natural Areas for Intensive Agriculture (ULSNA)

*Outline of the provisions and procedures*

**B.5.1**　In order to fill a gap in the coverage of projects that may be subject to the EIA process, a new consenting procedure was introduced on 4th February 2002, by the Scottish Ministers, for the use of uncultivated land and semi-natural areas for intensive agriculture (ULSNA). The process comprises a new regulatory procedure. In outline:

a. the Regulations make the Scottish Ministers (SEERAD) the Competent Authority;

b. the Regulations define a project involving the use of uncultivated land or semi-natural areas for intensive agricultural purposes (Reg. 3); and

c. prohibit such a project from being carried out without a screening opinion (Reg. 4);

d. if the screening opinion concludes that the project is likely to have a significant effect on the environment it becomes a relevant project (Reg. 5);

e. no relevant project can be undertaken without consent (Reg. 6);

f. the application for consent must be accompanied by an Environmental Statement which is then subject to publicity and consultation and taken into account by SEERAD before a consent can be granted (Reg. 9).

**B.5.2**　Thus, the EIA process is engaged for any project likely to have significant environmental effects that may involve the intensification of agriculture on uncultivated land or semi-natural areas. The Regulations go on to provide for the familiar steps in EIA including:

a. Statutory bodies must provide information to assist the preparation of the environmental statement (Reg. 8).

b. The applicant can ask for a scoping opinion from SEERAD as to the information to be provided in the Environmental Statement (Reg. 7).

c. SEERAD can require further information to be submitted (Reg. 10).

d. Statutory bodies must be consulted on the Environmental Statement and it must be publicised (Reg. 9).

e. There are provisions for transboundary effects (Reg. 11).

f. SEERAD must determine the application for consent taking account of the Environmental Statement and having regard to the requirements of the Habitats Regulations (Reg. 13).

## Section B.5 Part B

### Review of Old Mineral Permissions

**B.5.3**    Schedules 8, 9 and 10 of the *Town and Country Planning (Scotland) Act* 1997 require the review of old mineral permissions but did not contain any provisions for the requirement for EIA in respect of associated applications for the approval of conditions that would be made by operators to the planning authorities as part of the review procedure. In February 1999, the House of Lords ruled in *R v North Yorkshire County Council ex parte Brown and Cartwright (the Wensley Quarries case)* that the determination of a pre-1948 Interim Development Order application for new planning conditions constitutes development consent for the purposes of the EIA Directive, the effect of which was to require planning authorities to consider the need for EIA in such cases. The subsequent case of *R v Peak District National Park ex parte Bleaklow Industries Ltd* means that the Directive will also apply to the review of old permissions from 1948 to 1982 and subsequent periodic reviews of all mineral permissions. To accommodate this change, the Government introduced new Regulations in 2002.

**B.5.4**    Regulation 28(A(19)) of the Environmental Impact Assessment (Scotland) Review of Old Mineral Permissions Regulations (2002) (EIASROMPRO2) states that the deemed consent provisions of para 14(6)(b) of Schedule 8, para 9(8) of Schedule 9 and para 6(7) of Schedule 10 of the Town and Country Planning (Scotland) Act 1997 no longer apply where EIA may be required unless either the planning authority has adopted a screening opinion or the Scottish Ministers have made a Screening Direction to the effect that EIA is not required.

**B.5.5**    Regulation 28A(15–18) provides that where a planning authority requires an Environmental Statement they shall notify the operator and specify the date by which the statement is required. If on receipt of such a notification, the operator accepts that such an EIA is required, they must: write within 6 weeks or other agreed period from the date of notification stating that the operator accepts EIA is required and proposes to provide it by the specified date; and submit the statement and any documents required by Reg. 13 EIASR 99 by the specified date. If the operator disputes the need for EIA they must request within the 6 week period a screening direction from the Scottish Ministers.

**B.5.6**    The deemed approval of conditions is ended but there is a right of appeal against non-determination of applications for new conditions (Reg. 28A(23 and 24)).

**B.5.7**    With a few minor exceptions to adjust to the procedures for ROMPs, the provisions of the EIASR 99 apply to the cases where EIA is required for ROMPs, see further Annexe 2 Table 3 Part 1 below. SEDD Circular 1/2003 provides further guidance on the procedures.

## B.6 The Contents of an Environmental Statement
[See Annexes 1, 2 and 3]

**B.6.1** Paragraphs 64 onwards of Planning Advice Note (PAN) 58 and Regulation 2(1) and Schedule 4 Parts I and II of the EIASR 99, describe the contents of an 'Environmental Statement'. Essentially, an Environmental Statement is the written output of the developer's EIA team. It is intended to provide the focus for the EIA process by setting out all of the relevant information on which the public and consultees may then comment and which the Competent Authority must then take into account in making the decision.

**B.6.2** In the case of **Berkeley v SSETR**, the House of Lords commented that an Environmental Statement must not be a paper chase. Lord Hoffman said, *'the point about the Environmental Statement contemplated by the Directive is that it constitutes a single and accessible compilation, produced by the applicant at the very start of the application process, of the relevant environmental information and the summary in non-technical language.'*

**B.6.3** Its primary purpose, therefore, is to inform the decision maker of the environmental implications of the development. It should also inform statutory consultees, other interested bodies and members of the general public and provide a basis for consultation and debate.

★ Key information ★

**Box B.6.1**

An Environmental Statement should:

● be a 'stand-alone' and complete document (though not necessarily a single volume);

● provide enough detail to allow readers to form an independent judgement;

● be unbiased, neither advocating the project nor attempting to serve public relations purposes; and

● avoid technical discussion and terminology except where absolutely necessary.

**B.6.4** The EC Directive specifies, in Annexe III, and the EIASR 99 in Schedule 4 Part II, the information which **must** be included in an Environmental Statement. However, recognising that there may be occasions when some information may not be relevant to the consent procedure or may be impractical to collect, they also specify other information that an Environmental Statement may reasonably be required to include, by way of explanation or amplification (EIASR 99 Sch. 4 Part I) (see also Box B.6.3 below). The equivalent requirements in other Regulations are referred to in Annexe 2, Table 3 below.

**B.6.5** Thus, Regulation 2 of the EIASR 99 states that an Environmental Statement

*a. means a statement that includes such of the information referred to in Part I of Schedule 4 [Box B.6.3 below] as is reasonably required to assess the environmental effects of the development and which the applicant can, having regard in particular to current knowledge and methods of assessment, reasonably be required to compile, but*

*b. that includes at least the information referred to in Part II of Schedule 4 [Box B.6.2 below].*

★ Key information ★

**Box B.6.2**

**Information that MUST be included in an Environmental Statement – the minimum requirement.**

- A description of the development proposed, comprising information about the site and the design and size or scale of the development [EIASR 99 Sch. 4 Part II (1)]

- A description of the measures envisaged in order to avoid, reduce and, if possible, remedy significant adverse effects (the mitigating measures) [EIASR 99 Sch. 4 Part II (2)].

- The data required to identify and assess the main effects which that development is likely to have on the environment [EIASR 99 Sch. 4 Part II (3)].

- An outline of the main alternatives studied by the applicant or appellant and an indication of the main reasons for the choice, taking into account the environmental effects [EIASR 99 Sch. 4 Part II (4)].

- A non-technical summary of the above information [EIASR 99 Sch. 4 Part II (5)].

**B.6.6**    However, it should be noted that if matters in Box B.6.3 are not included in an Environmental Statement, but the Competent Authority decides that it is reasonably required to give proper consideration to the likely environmental effects of the proposed development, the Competent Authority can require the developer (by giving notice in writing) to submit the information specified in writing, but the Authority must have regard in particular to current knowledge and methods of assessment (Regulations 19, 36 and 60 EIASR 99).

**B.6.7**    The responsibility for carrying out the studies for the Environmental Statement and reporting the findings is placed on the developer although there are statutory responsibilities for public bodies to make available the relevant information which they hold. Some environmental issues, however, fall outside the competence or knowledge of any individual developer. In particular, the accurate characterisation of cumulative impacts of many developments in any one region or locality can rarely be satisfactorily assessed by individual developers. The regulations require the developer to include alternatives which have been considered; **if no alternatives have been considered none need be included in the Environmental Statement** (see further para B.6.9 below). An analysis of alternatives which, for example, involve different approaches to meeting social needs (rail travel instead of road, for example, or energy conservation instead of a new oil terminal) cannot reasonably be expected in a project Environmental Statement.

**B.6.8** Environmental Statements are increasingly available on CD or DVD and distribution in this form is compliant subject to the caveats explained in paragraphs D.1.8 and D.1.9 below.

★ **Key information** ★

### Box B.6.3

Matters normally to be included in an Environmental Statement which is reasonably required to assess the environmental effects of the development and which the applicant can reasonably be required to compile (subject to the minimum requirements listed in Box B.6.2 above).

● Description of the development  [EIASR 99 Sch. 4 Part I (1)]
A description of the development including, in particular:
a. the physical characteristics of the whole development, and the land use requirements during the construction and operational phases;
b. the main characteristics of the production processes, for instance, the nature and quantity of the materials to be used;
c. an estimate, by type and quantity, of expected residues and emissions (water, air and soil pollution, noise, vibration, light, heat, radiation etc.) resulting from the operation of the development.

● Alternatives considered  [EIASR 99 Sch. 4 Part I (2)]
An outline of the main alternatives studied by the applicant or appellant and an indication of the main reasons for choosing the development proposed, taking into account the environmental effects.

● Baseline environmental information  [EIASR 99 Sch. 4 Part I (3)]
A description of the aspects of the environment likely to be significantly affected by the development including, in particular, population, flora and fauna, soil, water, air, climatic factors, material assets, including the architectural and archaeological heritage, landscape and the inter-relationship between these factors.

● Environmental effects  [EIASR 99 Sch. 4 Part I (4)]
A description of the likely significant effects of the development on the environment which should cover the direct effects and any indirect, secondary, cumulative, short, medium and long-term, permanent and temporary, positive and negative effects of the development resulting from:
the existence of the development;
the use of natural resources;
any emission of pollutants, creation of nuisances, and elimination of waste;
and the description by the applicant or appellant of the forecasting methods used to assess the effects on the environment.

● Mitigation measures [EIASR 99 Sch. 4 Part I (5)]
A description of the measures envisaged to prevent, reduce and, where possible, to offset any significant adverse effects on the environment.

● A non-technical summary of the above information [EIASR 99 Sch. 4 Part I (6)]

● Technical difficulties and limitations [EIASR 99 Sch. 4 Part I (7)]
An indication of any difficulties, such as technical deficiencies or lack of know-how, encountered in compiling the required information.

## The Assessment of Alternatives

**B.6.9**    Circular 15/1999, Annexe C, paragraph 2 states that the Environmental Statement should contain *'An outline of the main alternatives studied by the applicant or appellant and an indication of the main reasons for his choice, taking into account the environmental effects'* This reflects the requirements of the Regulations and Directive (see Box B.6.3 above). The following points seem clear:

a. an applicant or appellant does not have to consider alternatives; but if they do

b. they must provide an 'outline of the main alternatives studied'; and

c. for each of the main alternatives studied, an indication of the main reasons for the choice, that is, why the alternative was not adopted, taking account of its environmental effects and those of the submitted project; noting that

d. the predicted environmental effects of the alternatives rejected may have been better or worse than the submitted project, so EIA does not absolutely constrain the selection of the submitted project in preference to alternatives studied, but it is reasonable to expect that a rational explanation would be included in the Environmental Statement as to why a more, or less, environmentally harmful project was chosen for submission.

## B.7    Importance of Compliance with the Directive

**B.7.1**    It is clear from this Handbook that the EIA process is complex and involves many decisions and judgements, all of which could be challenged by an aggrieved party either through the domestic Courts or by reference to the European Commission.

**B.7.2**    Failing to comply with the Regulations may make a decision to grant permission unlawful and lead to it being quashed by the Court. Although the Court has the power not to quash decisions where there has been procedural impropriety, this discretion is very limited in cases involving EIA because of the duty to comply with EC legislation. It can only be exercised where there has been **'substantial compliance'** with the Directive.

**B.7.3**    If the project is one to which the Regulations apply it is essential to comply fully with them. It is not sufficient to argue that EIA was not necessary because all of the information that could have been in the Environmental Statement was available elsewhere and was taken into account before the decision was taken; or that had an Environmental Statement been available the decision would have been the same.

**B.7.4**    In **Berkeley v SSETR**, the House of Lords unanimously emphasised the need to comply with the Regulations. It took the view that when considering compliance with the Regulations it was necessary to consider the EIA Directive. The Lords stressed that the importance of the EIA process extended beyond the decision on the application. Its purpose is to provide individual citizens with sufficient information about the possible effects and give them the opportunity to make representations. The Court was not entitled to decide after the decision had been made that the requirement of EIA could be dispensed with on the ground that the outcome would have been the same even if these procedures had been followed. In his leading judgement, Lord Hoffman noted that the Directive did not allow Member States to treat *'a disparate collection of documents produced by parties other than the developer and traceable only by a person with a good deal of energy and persistence as satisfying the requirement to make available to the public the information which should have been provided by the developer'.*

**B.7.5**    Individuals may complain to the European Commission that planning and other types of applications should have been subject to EIA, or that where EIA was undertaken the procedures were not followed correctly or the information in the Environmental Statement was inadequate. This can lead to formal legal proceedings between the Commission and the United Kingdom. This can be lengthy and prolonged and can increase uncertainty for developers and planning authorities.

**B.7.6**    Nothing can guarantee there will be no legal challenge. But all those involved in the EIA process can minimise the risk of such challenge being successful by ensuring compliance with all of the Regulations. For planning applications particularly:

- All applications should be properly screened and copies of screening opinions placed on the planning register.

- Environmental Statements should contain all of the information required by Schedule 4 of the Regulations.

- All of the significant effects that the project is likely to have on the environment should be identified and taken into account prior to a decision to allow the project to go ahead.

- The permission that is granted should relate only to the project whose environmental effects have been described, assessed and mitigated in the Environmental Statement.

- A record of all decisions and the reasons for them should be kept.

# Part C

# Prior to the Submission of the Environmental Statement

## C.1 Deciding whether EIA is Required: the 'Screening' Process

See also Section B.4 above and Annexe 4

| | Step in the EIA Process |
|---|---|
| **Stage 1:** **Before Submission of the Environmental Statement** | **Deciding whether EIA is required** |
| | Requiring submission of an Environmental Statement |
| | Preliminary contacts and liaison |
| | Scoping the Environmental Statement |
| | Information collection |
| | Describing baseline environmental information |
| | Predicting environmental impacts |
| | Assessing the significance of impacts |
| | Mitigation measures and enhancement |
| | Presenting environmental information in the Environmental Statement |
| **Stage 2:** **Submission of Environmental Statement and Consideration of Environmental Information** | Submission of Environmental Statement and project application for consent |
| | Consultation and publicity |
| | Requiring more information |
| | Negotiating modifications to the project |
| | Considering the environmental information |
| **Stage 3:** **Making the Decision** | Making the decision |
| | Guaranteeing compliance |
| **Stage 4:** **Implementation** | Implementation of mitigation and compensation measures |
| | Monitoring |
| | Review, reassessment and remedial measures |
| | Reporting |

**Table C.1.1 Summary of References for Equivalent Requirements in all EIA Regulations for Deciding whether EIA is Required**

| Topic | Regulations | Reference |
|---|---|---|
| Development requiring planning permission | EIASR 99 | Regs 4–6 |
| Development by a PA including local roads | EIASR 99 | Reg. 22 |
| Unauthorised development on appeal | EIASR 99 | Reg. 30–31 |
| Review of old mineral permissions | EIASROMPRO2/EIASR99 | Regs 4–6 and 28A |
| Motorways and trunk roads | EIASR 99 | N/A |
| Drainage improvements | EIASR 99 | Reg. 56 |
| Marine aquaculture | EIAFishFarmMWR 99 | Reg. 4 |
| Forestry works | EIAForestrySR 99 | Reg. 5–8 |
| Use of uncultivated land and semi-natural areas for agriculture | ULSNARO2 | Regs 4 + 5 |
| Irrigation, drainage and water management for agriculture | EIAWaterMRO3/EIASR99 | Regs 4–6 |
| Electricity power stations >50MW and overhead lines | ElecWorks EIASR 00 | Reg. 5 |
| Offshore electricity power stations >1MW | OffshoreGenStnsR02/ ElecWorks EIASR 00 | Reg. 5 |
| Gas pipelines not requiring planning permission | PGasTransPWEIAR 99 | Reg. 6 |
| Offshore oil and gas and pipelines | OffshorePPPAEER 99 | Reg. 6, 11 and 12 |
| Other pipelines | PipelineWEIAR 00 | Reg. 4 |
| Decommissioning nuclear installations | NuclearREIADR 99 | N/A |
| Harbours, docks, piers and ferries | HarbourWEIAR 99 | Reg. 4 & Sch. 3(5) HA 1964 |

**Note that the criteria and tests required to determine which projects are subject to EIA are set out in section B.3 and B.4, this section is about the procedures relating to the screening process.**

**C.1.2** Competent Authorities have a statutory duty to consider whether any project which they may be responsible for authorising is a project that should be subject to the EIA process. The developer can use statutory procedures to ask the Competent Authority or the Scottish Ministers whether an Environmental Statement will be required for a project. Guidance on this stage is also provided in PAN 58 at paragraphs 28–31 and 35–37 and in Circular 1/2003 paras 28–31 and Circular 3/2003 paras 8–9.

## Introduction

**C.1.2** Reference is made here to sections B.3 and B.4 and Annexe 4 which explain which projects require EIA. It is the responsibility of the competent authority to ensure that all relevant applications are 'screened' to establish whether EIA is required. In a planning authority, this will normally be carried out by the officer dealing with the planning application. But the decision is taken on behalf of the planning authority so it is important to ensure that the officers have delegated authority to do so. In **R v St Edmundsbury Borough Council ex parte Walton** a decision of the planning authority to grant planning permission was overturned because a decision not to require EIA was taken by an officer who had no formal delegation. PAN 58 gives best practice guidance advice in terms of the management of EIA applications.

**C.1.3** Where EIA is required, the authority must provide a written statement giving full reasons for its decision. There is no similar requirement where the authority decides that EIA is not required. However, it would be prudent for the authority to make and retain for its own use a clear record of the issues considered and the reason for its decision. This would be very useful in the event of any challenge to the planning decision based on EIA grounds (see B.7 above).

**C.1.4** If the project is EIA development the Competent Authority is prohibited from giving consent to the project until it has taken the environmental information into account (e.g. Reg. 3 EIASR 99) unless it is 'exempt development'. Exempt development is development which comprises or forms part of a project serving national defence purposes or a project in respect of which the Scottish Ministers have sent a copy of a direction to the relevant planning authority directing that the particular proposal is exempted from the application of the EIA regulations (Reg. 2 EIASR 99).

> **★ Key information ★**
>
> **Box C.1.1**
>
> The decision whether or not an EIA should be carried out for projects covered by the Regulations is a matter for the Competent Authority (e.g. the planning authority, Forestry Commission, the Scottish Ministers etc.).

### Developer's Options as to the Submission of an Environmental Statement

**C.1.5** If the project is a Schedule 1 project the developer has no option but to submit an Environmental Statement. However, if it is a Schedule 2 project, the developer has three courses of action. He can:

● Submit an Environmental Statement with the application for a consent, in which case the EIA process is initiated.

- Ask the Competent Authority for a screening opinion, which is a determination whether an Environmental Statement will be required.

- Submit an application without an Environmental Statement.

## Procedures for establishing whether or not EIA is required ('screening')

**C.1.6**   The determination of whether or not EIA is required for a particular development proposal can take place at a number of different stages:

a)  The developer may decide that EIA will be required and submit a statement which he refers to as an Environmental Statement for the purpose of the Regulations with the planning application.

b)  The developer may, before submitting any planning application, request a screening opinion from the planning authority. If the developer disputes the need for EIA (or a screening opinion is not adopted within the required period), the developer may apply to Scottish Ministers for a screening direction. Similar procedures apply to permitted development (see below).

c)  The planning authority may determine that EIA is required following receipt of a planning application. Again, if the developer disputes the need for EIA, the applicant may apply to Scottish Ministers for a screening direction.

d)  Scottish Ministers may determine that EIA is required for an application that has been called in for their determination or is before them on appeal.

e)  Scottish Ministers may direct that EIA is required at any stage prior to the granting of consent for particular development.

**C.1.7**   A developer may ask the Competent Authority for a screening opinion whether an Environmental Statement will be required before submitting the application (e.g. Reg. 5(1) EIASR99). The Competent Authority has 3 weeks (or such extended period as agreed between the parties) from receipt of the request in writing to provide its opinion, in writing (Reg. 5(4) EIASR99).

**C.1.8**   The Competent Authority may ask the developer for any additional information (Reg. 5(3) EIASR99) necessary to give an opinion, and may consult any of the statutory consultees (see section D.2 below) before giving their opinion. Where the Competent Authority decide whether or not an Environmental Statement is required and they adopt a screening opinion, they must notify the developer in writing  (Reg. 5(5) EIASR 99).

**C.1.9**   To avoid unnecessary delays it is important that every attempt should be made to issue screening opinions within the statutory 3 week period. The regulations do, however, allow for the authority and the applicant to agree a longer period. Unless there is such agreement, the authority has no legal authority to request EIA beyond the 3 week period. However, if it had not issued a screening opinion and it considered that EIA was required the authority could seek to persuade the applicant voluntarily to carry out an assessment and provide an Environmental Statement, which would be submitted in accordance with the Regulations. It can also request the Scottish Ministers to issue a screening direction to determine whether EIA is required.

**C.1.10** An authority can change its mind about a screening opinion, but should do so within the statutory period unless there is prior agreement of the applicant to extend the period. It is possible that additional information about the effects of the project not known to the authority when its screening opinion was given will come to light before a decision is taken on the application. If that information indicates that EIA is required the authority *must not* ignore it simply because it has already issued an opinion that EIA is not required. If the authority itself is unable to change its opinion, it should request a screening direction from the Scottish Ministers (who have a general power to direct whether EIA is required) before any decision is taken on the application.

**C.1.11** The case of **Fernback and Others v Harrow LBC** addressed this issue. In this case the Court held that a 'negative' screening opinion issued by a planning authority did not determine whether an application for planning permission was 'EIA Development' and a 'positive' one by the Planning Authority was determinative only in the absence of one by the Secretary of State (Scottish Ministers). On the other hand, an opinion by the Secretary of State, either way, is determinative.

**C.1.12** Failure by the Competent Authority to give an opinion in the three week period (or such extended period as agreed between the parties) means that the developer is entitled to request a screening direction from the Scottish Ministers. The developer may also request a screening direction from the Scottish Ministers where aggrieved by the decision of the Competent Authority to require EIA (Reg. 5(6) EIASR 99). A request for a screening direction can be made by the developer even if the Competent Authority required further information to be submitted and the information has not been submitted (this is in case the Competent Authority's requirements are unreasonable) (Reg. 5(7) EIASR 99).

**C.1.13** The Scottish Ministers have 3 weeks (or such extended period as agreed between the parties) within which to give notice in writing to the applicant of the screening direction. Such a Direction is final and the Scottish Ministers must inform the applicant and the competent authority of their decision (Reg. 7 EIASR 99).

**C.1.14** If a Competent Authority receives an application for consent, for a Schedule 1 or Schedule 2 project likely to have significant effects on the environment, it has three weeks (or such extended period as agreed between the parties) within which to give notice in writing to the applicant that an Environmental Statement should be submitted (Reg. 7 EIASR 99). In making this decision the Competent Authority may consult the statutory consultees (see section D.2 below).

**C.1.15** If the applicant receives from the Competent Authority a notice that in their opinion an Environmental Statement must be submitted, the applicant has three weeks in which to either:

a. confirm that a Statement will be submitted; or

b. unless the Scottish Ministers have already made a screening direction, inform the Competent Authority that the developer is writing to seek a screening direction from the Scottish Ministers (Regs 6 and 7 EIASR 99).

**C.1.16** If no Environmental Statement is submitted, or no request made to the Scottish Ministers for a screening direction, or the Scottish Ministers direct that an

**Figure 2**

The Procedure for Establishing whether an EIA is Required
NB This Figure is based on Figure 1, page 8 of Circular 15/1999

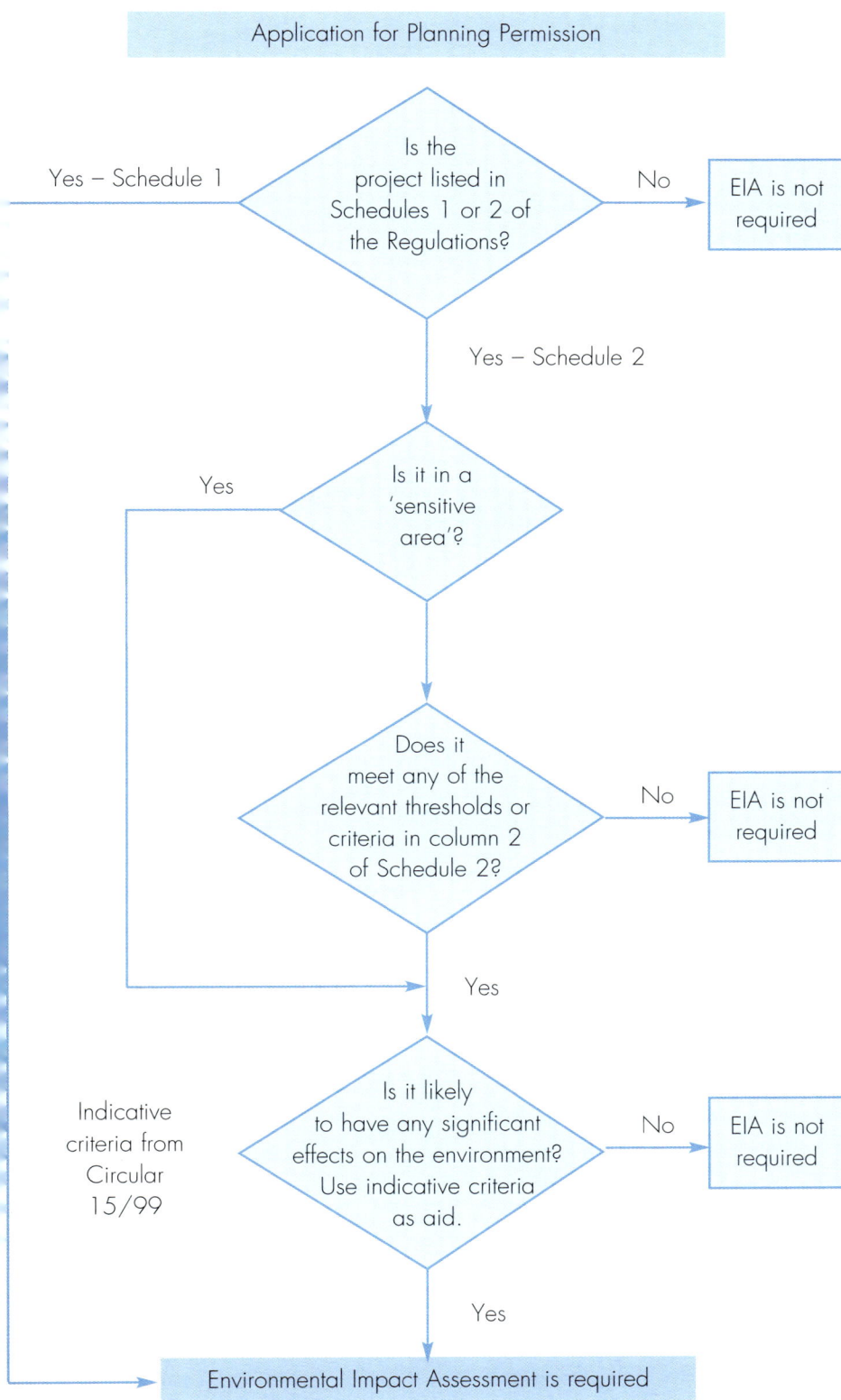

Application for Planning Permission

Yes – Schedule 1

Is the project listed in Schedules 1 or 2 of the Regulations?

No → EIA is not required

Yes – Schedule 2

Is it in a 'sensitive area'?

Yes

Does it meet any of the relevant thresholds or criteria in column 2 of Schedule 2?

No → EIA is not required

Yes

Is it likely to have any significant effects on the environment? Use indicative criteria as aid.

No → EIA is not required

Indicative criteria from Circular 15/99

Yes

Environmental Impact Assessment is required

Environmental Statement is required but none is submitted, the application is not actually invalid but consideration of the application is suspended until and unless an Environmental Statement is submitted. It would, however, be open to the Competent Authority to refuse permission on the grounds of inadequate information and, in any event, the Competent Authority should not grant any consent. (See Reg. 45 EIASR 99 and Circular 15/99 paragraph 50.)

**C.1.17** The procedure for establishing whether most development projects under the EIASR99 should be subject to EIA is shown in Figure 2 below.

## Screening Processes for Projects Using Uncultivated Land and Semi-Natural Areas for Intensive Agriculture (ULSNA)

**C.1.18** To avoid duplication, the ULSNA Regs do not apply to any project that is:

a. exempt by the Scottish Ministers or, in any event;

b. subject to the *Environmental Impact Assessment (Scotland) Regulations* 1999; or

c. an afforestation/woodland planting project described in Reg. 3(2) of the *EIA (Forestry) (Scotland) Regulations* 1999.

**C.1.19** Critical to the application of the ULSNA procedure is an understanding of the process for determination as to which plans and projects are a 'relevant project'. Firstly, the meaning of what constitutes a project needs to be understood. Reg. 2 defines a project very widely and should be interpreted widely to avoid any possibility of a breach of the requirements of the EIA Directive (see sections B.4 and B.7 above). A project means any intervention in the natural surroundings and landscape involving the use of uncultivated land or semi-natural areas for intensive agricultural purposes, including but not limited to carrying out construction works or installations or schemes.

**C.1.20** It is not always easy to define what may constitute intensive use in particular circumstances. For example, intervention in management by fertilising, reseeding or ploughing is likely to be clear in most cases, but other actions such as introduction of grazing, or increasing grazing levels, may be much more difficult to define as intensification. In this context, 'agricultural' purposes has the same meaning as the Agriculture (Scotland) Act 1948. However, 'intensive' is not defined, but for the purposes of the Regulations should generally be regarded as any change to the agricultural use or management of the land that would lead to any increase in production or output or anything else that may adversely affect the characteristics or extent of the uncultivated land or semi-natural habitats.

**C.1.21** Some kinds of construction works, installations and other schemes will constitute development requiring planning permission and all of those will be assessed under the *Environmental Impact Assessment (Scotland) Regulations* 1999. The ULSNA regulations apply to projects that do not require other forms of consent, for example, drainage, reseeding, ploughing etc.

**C.1.22** Reg. 4 requires that no person shall begin or carry out a 'project' without first obtaining a screening decision. The screening process is intended to determine which projects should be subject to EIA. All projects must be assessed. Those that are considered likely to have significant environmental effects are called 'relevant projects' and must be subject to an application for consent and thereby go through the EIA process. Those projects that will not be likely to have significant environmental effects do not need to be subject to an application for consent, will not require consent and do not need to be subject to the EIA process.

**C.1.23**   Reg. 5 and procedures established by SEERAD mean that anyone proposing to undertake a project, as defined, must submit to SEERAD a completed pro-forma including details of the location (on a plan), nature, extent and purpose of the project and its possible effects on the environment and such other information or representations as the applicant may wish to make (Reg. 5(1)). It should be noted, however, that in respect of the nature, extent and purpose of the project and its possible effects on the environment, the regulations require only a 'brief' description. It is not a requirement to submit an Environmental Statement at this stage. SEERAD can ask for further information only to the extent necessary to make a screening opinion, not necessarily that required to determine an application for consent (see also section B.4 above)

**C.1.24**   The criteria for deciding whether any project subject to a screening application is likely to have significant environmental effects are set out in Schedule 1 of the Regulations. Any project which SEERAD considers to be likely to have a significant effect on a Natura 2000 site shall automatically be subject to the consent procedure and must be subject to EIA.

**C.1.25**   SEERAD has 35 days in which to decide whether the project being screened is a 'relevant project' (i.e. one that will be likely to have significant environmental effects). The 35 days runs from 'the notified date' (please see below for a definition). Upon making the decision SEERAD must notify the applicant and consultees who might wish to be informed and enter the decision in a public register. The decision must include full reasons. If an applicant has not received a decision in the 35 day period, or longer period agreed with SEERAD, the project is deemed to be a relevant project (subject to consent and EIA), unless and until SEERAD issues a screening decision to the contrary.

**C.1.26**   A screening decision is valid for 3 years after which time a project would need to be resubmitted for screening. The 3 year period runs from 'the notified date' which is

a.  the date SEERAD notifies the applicant that they received the application; or

b.  the date that SEERAD required further information to be provided; or

c.  such date as may be agreed between SEERAD and the applicant.

## Permitted Development

**C.1.27**   The *Town and Country Planning (General Permitted Development) (Scotland) Order* 1992 (GPDO) grants a general planning permission (usually referred to as permitted development rights) for various specified types of minor, non-contentious developments, or developments that need another regulatory consent, the procedures for which would merely duplicate the planning process (e.g. Land Drainage Consents and Harbour Revision Orders). The majority of permitted developments are very unlikely to fall within any of the descriptions in Schedules 1 or 2, but it is possible that some might, for example a large scale water management scheme for agriculture.

**C.1.28** The provisions of the Permitted Development Order (insofar as they relate to Schedule 1 or Schedule 2 development) are amended by regulation 47(4) EIASR 99 so that:

a. Schedule 1 development is not permitted development. Such developments always require the submission of a planning application and an Environmental Statement.

b. Schedule 2 development does not constitute permitted development unless the planning authority has adopted a screening opinion to the effect that EIA is not required. Where the authority's opinion is that EIA is required, permitted development rights are withdrawn and a planning application must be submitted and accompanied by an Environmental Statement.

These requirements do not apply to certain types of permitted development, described below in paragraphs C.1.30–32.

**C.1.29** A request for a screening opinion in relation to permitted development should be made in accordance with the provisions which apply to requests for a pre-application screening opinion set out in Reg. 5 EIASR99 and paras C.1.6 to C.1.17 above. There are similar rights to request Scottish Ministers to make a screening direction if a developer disagrees with an opinion that EIA is required, or where the planning authority fails to adopt any opinion within 3 weeks (or such longer period as is agreed in writing). Such requests should be made in accordance with the procedures in Reg. 6 EIASR 99. Requests to the planning authority for a screening opinion can be made alongside any 'prior notification' which may be required in relation to any particular form of permitted development.

## Permitted Development (exceptions to the Town and Country Planning EIA Provisions)

**C.1.30** The provisions described in paragraphs C.1.27 to C.1.29 above do not apply to the following permitted developments because these are either exempted by the Directive or subject to other consenting procedures to which other EIA regulations, or other parts of the EIASR 99 apply (Regulation 47(3), (4), (5), (6) EIASR 99):

a) Part 7 forestry buildings and operations (because they are subject to the Forestry EIA Regulations (EIAForestrySR99).

b) Class 26 of Part 8 development comprising deposit of waste material resulting from an industrial process (excluded because it concerns projects begun before the date on which the Directive came into operation).

c) Part 11 development under local or private acts or orders (being exempt as described in paragraph B.3.10 above).

d) Class 39(1)(a) of Part 13 development by public gas transporters (because they are subject to the Gas Transporters EIA Regulations (PGasTransPWEIAR 99).

e) Class 58 of Part 17 development by licensees of the Coal Authority (because it concerns projects begun before the date on which the Directive came into operation).

f)   Class 64 of Part 18 deposit of mining waste (because it concerns projects begun before the date on which the Directive came into operation).

g)  Class 20 Part 6 land drainage development that is subject to the Land Drainage EIA procedures under Part IV of the EIASR 99 (because they have their own EIA procedure in Part IV).

**C.1.31**   Certain developments permitted by classes 54, 59, 60 and 63 (certain types of mineral and mineral waste operations) and begun before 1 August 1999 are also excluded, but these provisions are complex and you will need specialist advice on these rare cases.

**C.1.32**   Development which comprises or forms part of a project serving national defence purposes is excluded by virtue of Article 1.4 of the Directive (see definition of 'exempt development' in the glossary below (regulation 2(1) and at C.1.4.).

## C.2 Requiring the submission of an Environmental Statement

| Step in the EIA Process | |
|---|---|
| **Stage 1:**<br>**Before Submission of the**<br>**Environmental Statement** | Deciding whether EIA is required |
| | **Requiring submission of an Environmental Statement** |
| | Preliminary contacts and liaison |
| | Scoping the Environmental Statement |
| | Information collection |
| | Describing baseline environmental information |
| | Predicting environmental impacts |
| | Assessing the significance of impacts |
| | Mitigation measures and enhancement |
| | Presenting environmental information in the Environmental Statement |
| **Stage 2:**<br>**Submission of Environmental**<br>**Statement and Consideration of**<br>**Environmental Information** | Submission of Environmental Statement and project application for consent |
| | Consultation and publicity |
| | Requiring more information |
| | Negotiating modifications to the project |
| | Considering the environmental information |
| **Stage 3:**<br>**Making the Decision** | Making the decision |
| | Guaranteeing compliance |
| **Stage 4:**<br>**Implementation** | Implementation of mitigation and compensation measures |
| | Monitoring |
| | Review, reassessment and remedial measures |
| | Reporting |

**Table C.2.1 Summary of References for Equivalent Requirements in all EIA Regulations for Requiring Submission of an Environmental Statement**

| Topic | Regulations | Reference |
|---|---|---|
| Development requiring planning permission | EIASR 99 | Regs 3, 7 + 9 |
| Development by a PA including local roads | EIASR 99 | Reg. 22 |
| Unauthorised development on appeal | EIASR 99 | Regs 29, 33–34 |
| Review of old mineral permissions | EIASROMPRO2/EIASR99 | Regs 3–9 + 28A |
| Motorways and trunk roads | EIASR 99 | N/A |
| Drainage improvements | EIASR 99 | Reg. 57 |
| Marine aquaculture | EIAFishFarmMWR 99 | Reg. 3 + 5 |
| Forestry works | EIAForestrySR 99 | Reg. 4 |
| Use of uncultivated land and semi-natural areas for agriculture | ULSNARO2 | Regs 6 + 9 |
| Irrigation, drainage and water management for agriculture | EIAWaterMRO3/EIASR99 | Regs 3, 7 + 9 |
| Electricity power stations >50MW and overhead lines | ElecWorks EIASR 00 | Regs 3, 4 + 6 |
| Offshore electricity power stations >1MW | OffshoreGenStnsRO2/ElecWorks EIASR 00 | Regs 3, 4 + 6 |
| Gas pipelines not requiring planning permission | PGasTransPWEIAR 99 | Reg. 3 |
| Offshore oil and gas and pipelines | OffshorePPPAEER 99 | Regs 4, 5 + 11 |
| Other pipelines | PipelineWEIAR 00 | Regs 3, 11–13 |
| Decommissioning nuclear installations | NuclearREIADR 99 | Regs 3–5, 8 +16 |
| Harbours, docks, piers and ferries | HarbourWEIAR 99 | Regs 11–14 |

**[See also Figure 2, Sections B.3, C.1, C.3, Annexes 4 and 7]**

## Statutory Provisions and Government Guidance

**C.2.1**    Competent Authorities and the Scottish Ministers have a statutory power to require submission of an Environmental Statement in a particular case and a statutory duty not to grant any form of consent to a project which should be subject to the EIA process, without considering the environmental information. The Ministers have wide powers to enforce the EIA regime in Scotland  (see paras 78–79 Circular 15/1999).

## Introduction

**C.2.2**    Reference is made to section C.1 above and to section B.3 and Annexe 4 which explain which projects require EIA.

**C.2.3**    Whether or not it is consulted about the need for EIA, a consultee in the EIA process may decide independently to advise the Competent Authority that it considers that an EIA should be carried out when it receives an application for comment as part of the regular consultation process. In this case, the consultee would have to advise the Competent Authority in sufficient time to allow it to reach a decision and advise the developer accordingly within the 3 week period (Regs 7 and 20 EIASR 99 and paragraph 67 Circular 15/1999).

**C.2.4**    If the Competent Authority decides that it does not wish to follow the consultee's advice in a particular case, then the consultee can ask the Scottish Ministers to issue a Direction to require EIA to the Competent Authority under the Regulations (EIASR 99 Reg. 4 and Article 16 of the GDPO and paras 49 and 78 Circular 15/1999).

**C.2.5**    It should be noted, however, that the Scottish Ministers do not have to wait for a developer or a Competent Authority to ask for a Direction. They can act at any time. (See Circular 15/1999, paragraphs 49 and 78.)

## Projects Using Uncultivated Land and Semi-Natural Areas for Intensive Agriculture (ULSNA)

**C.2.6**    Reg. 6 requires that no person shall begin or carry out a relevant project (see the screening process in C.1 above) without first obtaining SEERAD consent. Reg. 9 explicitly requires an application for consent to include the submission of an Environmental Statement.

**C.2.7**    Thus, all relevant projects are all subject to EIA and must all be submitted with an Environmental Statement. No application for consent under these regulations would be valid without an Environmental Statement.

## C.3    Preliminary Contact and Liaison

| | Step in the EIA Process |
|---|---|
| **Stage 1:**<br>**Before Submission of the**<br>**Environmental Statement** | Deciding whether EIA is required |
| | Requiring submission of an Environmental Statement |
| | **Preliminary contacts and liaison** |
| | Scoping the Environmental Statement |
| | Information collection |
| | Describing baseline environmental information |
| | Predicting environmental Impacts |
| | Assessing the significance of impacts |
| | Mitigation measures and enhancement |
| | Presenting environmental information in the Environmental Statement |
| **Stage 2:**<br>**Submission of Environmental**<br>**Statement and Consideration of**<br>**Environmental Information** | Submission of Environmental Statement and project application for consent |
| | Consultation and publicity |
| | Requiring more information |
| | Negotiating modifications to the project |
| | Considering the environmental information |
| **Stage 3:**<br>**Making the Decision** | Making the decision |
| | Guaranteeing compliance |
| **Stage 4:**<br>**Implementation** | Implementation of mitigation and compensation measures |
| | Monitoring |
| | Review, reassessment and remedial measures |
| | Reporting |

**[See also Figure 2, Sections C.1 and C.4]**

## Statutory Provisions and Government Guidance

**C.3.1**    Preliminary contacts and liaison are non-statutory procedures. Guidance on this stage is  provided in PAN 58 at paragraphs 32–34.

## Advantages

**C.3.2**    Early contact and liaison about EIA cases is of benefit to the project proposers and to SNH. It should help to reduce SNH's time input later in the process and increase the account taken of natural heritage issues in the Environmental Statement. The preparation of the Statement is the duty of the project proposer.

**C.3.3**    The whole EIA process should be carefully planned and programmed by all of the participants. The developer should consider the resources required and appoint a coordinator with overall responsibility for the coordination and production of the Environmental Statement and its submission. The coordinator should assemble a team with the right experience and expertise. The developer should also allow sufficient time for the assessment to be conducted properly and as thoroughly as necessary. The advice of the main parties in the EIA process should be sought at as early a stage as possible. Preliminary dialogue can be of great assistance to the developer, in understanding the potential concerns, and for the Competent Authority and key consultees in understanding the project and steering the preparation of the Environmental Statement.

**C.3.4**   One of the important contributors to the success of an Environmental Statement can be the extent of consultation prior to its submission and the careful consideration of its scope and content at the very beginning of the process (see Section C.4 below). The issue of drafts or draft extracts of the Environmental Statement, to key consultees and the Competent Authority, before the submission of the final statement and before the design is finalised, can improve the Environmental Statement considerably and expedite the EIA and decision making processes.

**C.3.5**   It is increasingly likely that consultees will be informed or consulted about a project when it is in its very early stages. This can be frustrating because there may be little information about it. Consultees can also feel cautious about commenting on a proposal before its full implications can be ascertained.

## C.4    Scoping the Environmental Statement

| | Step in the EIA Process |
|---|---|
| **Stage 1:** **Before Submission of the Environmental Statement** | Deciding whether EIA is required |
| | Requiring submission of an Environmental Statement |
| | Preliminary contacts and liaison |
| | **Scoping the Environmental Statement** |
| | Information collection |
| | Describing baseline environmental information |
| | Predicting environmental impacts |
| | Assessing the significance of impacts |
| | Mitigation measures and enhancement |
| | Presenting environmental information in the Environmental Statement |
| **Stage 2:** **Submission of Environmental Statement and Consideration of Environmental Information** | Submission of Environmental Statement and project application for consent |
| | Consultation and publicity |
| | Requiring more information |
| | Negotiating modifications to the project |
| | Considering the environmental information |
| **Stage 3:** **Making the Decision** | Making the decision |
| | Guaranteeing compliance |
| **Stage 4:** **Implementation** | Implementation of mitigation and compensation measures |
| | Monitoring |
| | Review, reassessment and remedial measures |
| | Reporting |

**Table C.2.1 Summary of References for Equivalent Requirements in all EIA Regulations For Requiring Submission of an Environmental Statement**

| Topic | Regulations | Reference |
|---|---|---|
| Development requiring planning permission | EIASR 99 | Reg. 10–11 |
| Development by a PA including local roads | EIASR 99 | N/A |
| Unauthorised development on appeal | EIASR 99 | N/A |
| Review of old mineral permissions | EIASROMPRO2/EIASR99 | Regs 10–11 and 28A |
| Motorways and trunk roads | EIASR 99 | N/A |
| Drainage improvements | EIASR 99 | N/A |
| Marine aquaculture | EIAFishFarmMWR 99 | Reg. 6 |
| Forestry works | EIAForestrySR 99 | Reg. 9 |
| Use of uncultivated land and semi-natural areas for agriculture | ULSNARO2 | Reg. 7 |
| Irrigation, drainage and water management for agriculture | EIAWaterMRO3/EIASR99 | Regs 10–11 |
| Electricity power stations >50MW and overhead lines | ElecWorks EIASR 00 | Reg. 7 |
| Offshore electricity power stations >1MW | OffshoreGenStnsRO2/ ElecWorks EIASR 00 | Reg. 7 |
| Gas pipelines not requiring planning permission | PGasTransPWEIAR 99 | Reg. 7 |
| Offshore oil and gas and pipelines | OffshorePPPAEER 99 | Reg. 7 |
| Other pipelines | PipelineWEIAR 00 | Reg. 5 |
| Decommissioning nuclear installations | NuclearREIADR 99 | Reg. 6 |
| Harbours, docks, piers and ferries | HarbourWEIAR 99 | Regs 4 and Sch. 3(6) HA 1964 |

**[See also Figure 2, Section C.3,  Appendices 1–6 and the Scoping Guide in Appendix 6]**

## Statutory Provisions and Government Guidance

**C.4.1**    Scoping of an Environmental Statement is a statutory procedure whenever requested by the applicant. That is, before making an application, an applicant may ask the  Competent Authority for their formal opinion on the information to be supplied in the Environmental Statement (a 'scoping opinion'). This provision allows the developer to be clear about what the authority considers the main effects of the development are likely to be and, therefore, the topics on which the Environmental Statement should focus (EIASR 99 Reg. 10).

**C.4.2**    The developer must include the same information as would be required to accompany a request for a screening opinion and both requests may be made at the same time (EIASR 99 Reg. 10(2) and (5)). An applicant may also wish to submit a draft outline of the Environmental Statement, giving an indication of what he considers to be the main issues, to provide a focus for the authority's considerations. The authority can require the applicant to submit any further information needed to adopt a scoping opinion. The authority must consult the consultation bodies and the developer before adopting its scoping opinion.

**C.4.3**    The planning authority must adopt a scoping opinion within 5 weeks of receiving a request (or, where relevant, of adopting a screening opinion–EIASR 99 Reg. 10(5)). This period may be extended by agreement in writing. As a starting point, authorities should study the definition of Environmental Statement in EASR Reg. 2(1) and Schedule 4 and the guidance elsewhere in Circular 15/1999 paras 82–86 and 90–97 and Annexe A). In addition, authorities may find it useful to consult other published guidance, such as the European Commission, DG XI.B.2, May 1996, *EIA–Guidance on Scoping*. PAN 58 also refers to the scoping process at paras 40–43. The Department of Environment (England) publication of 1995 *A Good Practice Guide to the Preparation of Environmental Statements* may also be useful.

**C.4.4**    The scoping opinion must be kept available for public inspection for 2 years (with the request and documents submitted by the applicant as part of that request) at the place where the planning or other type of register is kept. If a planning application is subsequently made for development to which the scoping opinion relates, the opinion and related documents should be transferred to Part 1 of the register with the application (EIASR 99 Reg. 20).

**C.4.5**    There is no provision to refer a disagreement between the developer and the Competent Authority over the content of an Environmental Statement to Scottish Ministers (although on call-in or appeal Scottish Ministers will need to form their own opinion on the matter). However, where a Competent Authority fails to adopt a scoping opinion within 5 weeks (or any agreed extension), the developer may apply to the Scottish Ministers for a **scoping direction** (EIASR Reg. 10(7)). This application must be accompanied by all the previous documents relating to the request for a scoping opinion, together with any additional representations that the applicant wishes to make. The applicant should also send a copy of the request and any representations to the Competent Authority, who are free to make their own additional representations.

**C.4.6**    Under Reg. 11 of the EIASR 99 Scottish Ministers must make a scoping direction within 5 weeks from the date of receipt of a request, or such longer period as they may reasonably require. They **must** consult the consultation bodies and the developer beforehand. Copies of the scoping direction will be sent to the developer and to the Competent Authority, which must ensure that a copy is made available for inspection with the other documents referred to in C.4.5.

## Effect of a scoping opinion or direction

**C.4.7**  An Environmental Statement is not necessarily invalid if it does not fully comply with the scoping opinion or direction. However, as these documents represent the considered view of the Competent Authority or Scottish Ministers, a statement that does not cover all the matters specified in the scoping opinion or direction will probably be subject to calls for further information under Reg. 19 (see D.6 below).

**C.4.8**  The fact that a Competent Authority or the Scottish Ministers have given a scoping opinion or scoping direction does not prevent them from requesting further information at a later stage under Reg. 19 EIASR 99. Where Scottish Ministers have made a scoping direction in default of the Competent Authority, the authority must still take into account all the information they consider relevant. In practice there should rarely be any difference between the relevant information and that specified by Scottish Ministers.

**C.4.9**  PAN 58 states at paras 41–42 that:

*The purpose of scoping is:*

- *to focus the EIA on the environmental issues and potential impacts which need the most thorough attention;*

- *to identify those which are unlikely to need detailed study;*

- *to provide a means to discuss methods of impact assessment and reach agreement on the most appropriate.*

By drawing on the knowledge of the planning authority and consultees, a scoping exercise will help the developer to identify the main issues quickly. It also gives an early indication of where mitigation measures may be necessary and should help to reduce requests for further information once the Environmental Statement is submitted. In some cases developers have used a forum of interested parties to discuss the issues informally prior to the formal scoping stage. The matters identified by the scoping exercise will derive from the nature of the project, the site and the environment.

**C.4.10**  The PAN goes on to say (para 40) that:

*For the planning authority in particular, this is an opportunity to act positively and provide early advice on the EIA process, methodologies, sensitive issues and sources of information. Early involvement of all parties is encouraged. … The applicant has first to provide information on the proposal including a site plan, a brief description of the proposal and its possible effects. At project initiation stage developers may wish to carry out scoping to a limited extent, possibly on a confidential basis, prior to seeking the formal opinion of the planning authority. The scoping is a key part of the EIA process but additional issues may still emerge as work progresses and the planning authority is not precluded from requiring the applicant to submit further information at a later stage.*

**C.4.11** Some Environmental Statements have contained excessive detail relating to issues that are irrelevant or of little importance to the decision. Others have overlooked issues which, when they came to light later in the process, proved to be decisive in the decision. Developers should not have to pay the cost and experience the delay involved in addressing issues that are obviously not significant. Competent Authorities, consultees and the public should not have to deal with large volumes of material which is irrelevant to the decision to be made.

★ **Key information** ★

**Box C.4.1**

**Advantages of Scoping**

The 'scoping' of the Environmental Statement can avoid excessive detail and omission of important issues and help the EIA process to focus on key issues. It is an important contribution to the EIA process. The Competent Authority has a statutory duty to provide a scoping opinion.

## Objects of Scoping

**C.4.12** Scoping should:

- identify the most important environmental effects and agree that these will be dealt with thoroughly;

- for these effects, agree a common basis for the survey, analysis and assessment methods and how information about effects and related issues should be presented;

- agree what other potential effects may be significant and ensure that they are investigated and assessed as far as may be necessary;

- agree which potential effects and issues are not likely to be significant and indicate that these will be listed and explained in the Environmental Statement, but not covered in detail; and

- identify alternative solutions and options to be examined to see whether they would have greater or lesser, or different environmental effects.

★ **Good EIA practice** ★

**Box C.4.2**

Developers or their consultants should:

- Make early site visits in order to ensure that matters of natural heritage and other environmental concern are identified at an early stage.

- Establish appropriate consultation arrangements with interested parties including the Competent Authority.

- Conduct the scoping exercise in a systematic manner using scoping matrices and producing a Scoping Report where appropriate.

- Agree baseline survey requirements, prediction methods and evaluation criteria with appropriate bodies, including non-governmental bodies where they have expertise.

## Products of Scoping

**C.4.13** The scoping exercise should provide three principal products:

a. A list of activities which may cause environmental effects, together with initial estimates of their likelihood and their potential magnitude.

b. A list of natural heritage receptors that are likely to be affected by the different stages or activities of the project.

With (a) and (b) combined into a scoping matrix:
c. A plan for conducting the technical studies, including details of methods to be used and resources required.

**C.4.14** The findings of the scoping process should be formally presented in the form of a **Scoping Report**, with a scoping matrix, although the production of such a report is not a requirement of the EIA Regulations. The Regulations do not specify what form the scoping opinion should take.

**C.4.15** A scoping report provides the developer with a valuable check on the progress and competence of the EIA team, and provides an opportunity for interested parties to comment on the proposed coverage and methodology of the Environmental Statement. Since the scoping of the assessment should also be reported within the Environmental Statement the effort expended in producing the report will not be wasted.

**C.4.16** Ultimately, the Environmental Statement should contain detailed descriptions of the scoping process, including a list of all consultees involved, the concerns raised by those consultees, copies of scoping letters and minutes of meetings held.

**C.4.17** However, scoping should not be a formality, simply because it is seen as a good thing; it should be acted upon by the developer. Research (25) showed that:

*involvement of conservation organisations at the scoping stage did not necessarily lead to detailed consideration of ecological issues in the resulting environmental statement. In one notable example for a development within a site of considerable ecological value, there was direct reference to a letter seeking advice from the then Nature Conservancy Council and also of the NCC's reply. That advice, including the suggestion that an ecologist be employed to conduct the ecological assessment, was not acted upon and the resulting environmental statement, particularly the ecological section, was extremely weak.... The scoping process has been wasted.*

**C.4.18** Developers are encouraged to use the published best practice guidance (21)(22) and the further guidance in the Appendices of this Handbook. The Attachment to this Handbook provides a Review Package which includes a guide through the scoping stage.

### Scoping Projects Using Uncultivated Land and Semi-Natural Areas for Intensive Agriculture (ULSNA)

**C.4.19** Reg. 7 ULSNARO2 provides for anyone needing to apply for consent (and required to submit an environmental statement) for a relevant project to apply to SEERAD for a 'scoping opinion'. The request is optional and need not be made if the applicant so decides. If a request is made it must be before the application for consent is submitted. SEERAD has 35 days to provide the scoping opinion. SEERAD may consult consultation bodies and the applicant in respect of a scoping opinion. SEERAD may require further information from the applicant if that submitted is insufficient to provide a scoping opinion; the applicant would have 28 days in which to provide it.

## Scope of Impacts Covered

**C.4.20**  Appendices 1 to 6 of this Handbook give many examples of potential impacts that may be considered when drawing up scoping matrices. A comprehensive list is not possible to draw up owing to the diversity of projects likely to arise. The examples in the Appendices should be adapted in every case and each Environmental Statement will require its own impact matrix to be developed. This is a task for the developer's project team. However, consultees should be asked to comment on the scoping matrix and to receive drafts and a final version.

**C.4.21**  When commenting, it is particularly important to bear in mind the different stages in the life of a project. Often an Environmental Statement will concentrate on operational stages, some will include construction and/or restoration stages, but few will include all the stages of a development unless prompted to do so. The main stages are summarised below in Figure 3. Not every project will go through every stage. Some projects, such as minerals and waste disposal, will have several stages present on the site at the same time, at some stages in the project life, e.g. site preparation, extraction, restoration and after care. Each stage can be subject to phases.

**C.4.22**  The impacts of associated infrastructure that will be essential for a project's operation should be covered (e.g. grid connections from an electricity generator). The impacts of new developments which are likely to follow on from the project in question should also be considered (e.g. a new runway following the development of a new terminal at an existing airport). However, at the end of the day, the Competent Authority cannot insist that more is covered in the Environmental Statement than is the result of the development proposal subject to their consent, see para B.4.26–27, and para 4.6 of Circular 15/1999 below.

**C.4.23**  Direct and indirect impacts that arise from the use of natural resources for the project may be included in the Environmental Statement, by way of explanation or amplification. They should, therefore, be considered in the EIA where significant. However, because these effects are included in Schedule 4 Part I of the EIASR 99, and not in Schedule 4 Part II, the Competent Authority can only require these details to be submitted in the Environmental Statement where they decide that the information is reasonably required to give proper consideration to the likely environmental effects of the proposed development and the applicant can, having regard to current knowledge and methods of assessment, reasonably provide it. In these cases, the Competent Authority can require the developer (by giving notice in writing) to submit the information specified in writing.

**C.4.24**  There may be some debate as to whether a particular development will cause indirect impacts of significance on natural resources. The key question is whether the new development will alter demand for the raw materials to the extent that significant impacts may be caused by the need to supply these. The following examples illustrate the point:

- A major road may have a substantial requirement for aggregates that would have to be extracted from local sources because of the high transport cost of these materials. The related impacts should be addressed in the Environmental Statement.

- It would be reasonable to consider the impact of gas extraction for a new turbine if the demand for gas created by the turbine would lead to an increased rate of extraction such that additional infrastructure would be needed.

- An EIA might consider the impacts of North Sea capelin exploitation if that was the main natural resource for a fish meal factory.

## Scoping an Outline Planning Application

**C.4.25**   Reference is made to sections C.1, D.6 and D.11 of this Handbook which set out further advice on outline planning applications and the powers that a planning authority has in respect of requiring more information to be submitted in respect of outline planning applications, and also refer to important case law.

**C.4.26**   Circular 15/1999, paragraph 48 provides the following advice on outline applications as follows:

*Where EIA is required for a planning application made in outline, the requirements of the Regulations must be fully met at the outline stage since reserved matters cannot be subject to EIA. When any planning application is made in outline, the planning authority will need to satisfy themselves that they have sufficient information available on the environmental effects of the proposal to enable them to determine whether or not planning permission should be granted in principle. In cases where the Regulations require more information on the environmental effects for the Environmental Statement than has been provided in an outline application, authorities should request further information under regulation 19. This may also constitute a request under article 4(3) of the GDPO.*

## Relevance of Natural Heritage Designations to EIA Scoping

**C.4.27**   Whether or not an Environmental Statement has been required because of the project's impacts on a designated area, all natural heritage and other designations affected should always be addressed in an Environmental Statement. It is important, therefore, at the scoping stage, to ensure that the developer is aware of and understands the significance and purpose of all relevant designations. The effects on the designation should be carefully assessed and conclusions drawn as to their significance. This should include reference to national, development plan and other policies relating to the designations.

**C.4.28**   The designations in box C.4.3 below should be included in the Environmental Statement wherever applicable.

### Selecting Methods for Impact Assessment

**C.4.29**  Whilst there can be no standard form of method for assessing the wide variety of impacts that may be encountered in an Environmental Assessment, the criteria in Box C.4.4 may assist in considering the selection of methods on a case-by-case basis, particularly for ecological and geological assessments. Reference should also be made to Appendices 1–6 of this Handbook.

**C.4.30**  There should generally be less need to depart from the well defined procedures set out in Appendix 1, for landscape and visual impact assessment, although even here there may be a need to consider the most appropriate form of visualisation, e.g. by way of computer generated photomontage, artist's impressions or computer generated visual envelopes and zones of visual influence.

**Figure 3  Project Life Stages**

| Overall Stages | Life Stage | Examples of Sources of Potential Effects |
|---|---|---|
| Pre-consent Stages | Site Finding | Potential changes in management or use of sites with potential, effects of neglect and blight. |
| | Site Investigation/ Exploration | Physical impacts to site from equipment for drilling and testing, anemometers and other testing and sampling equipment. |
| | Environmental Surveys | Disturbance and other impacts resulting from natural heritage, archaeological and other sampling and surveys. |
| Pre-construction Stages | Site Acquisition/ Requisitioning | Abandonment of normal land use or management whilst site acquired, neglect or removal of assets, e.g. trees for timber value. Fencing may change. |
| | Advance Mitigation | Earth moving, planting and other mitigation works in advance of commencement of main construction. |
| | Site Preparation | Permanent and temporary landtake, earth moving, soil stripping, overburden removal, removal of site features, access, water abstraction and drainage works, fencing. |
| Construction Stages | Construction | Storage and handling of materials, construction activities, earth moving, soil and subsoil compaction and stripping, blasting, drilling, piling, water abstraction and drainage works, tunnels, culverts, access by vehicles plant and equipment, compounds, parking, accidental spillages, noise, vibration, light, disruption to public access. |
| | Restoration of Construction Works | Translocation from other sites, seeding, turfing, planting and cultivating. Compounds, use of plant and equipment, vehicular access, storage of materials, movement, soil and sub soil handling, testing and site investigations/surveys. |
| | Commissioning | Testing, repairing, altering, moving and otherwise modifying project, often at short notice. |

| Overall Stages | Life Stage | Examples of Sources of Potential Effects |
|---|---|---|
| Operational Stages | Operational Phase | Gaseous and particulate emissions, noise, vibration, disturbance, effluents, light, water abstraction and discharges, vehicular access and parking. |
| | Monitoring | Monitoring investigations, surveys etc., repair, maintenance, replacement, emergencies (foreseen and unforeseen), increased maintenance and repair as project ages. |
| Decommissioning and Restoration | Decommissioning | Run-down in outputs, changes in balance of emissions and effluents, changes in noise and disturbance, light, water abstraction and discharges, fluctuations in outputs and activity. |
| | Demolition/ Removal | Storage and handling of materials, demolition activities, earth moving, soil compaction, blasting, drilling, water abstraction and drainage works, tunnels, culverts, access by vehicles plant and equipment, compounds, parking, accidental spillages, noise, vibration, light, disruption to public access. |
| | Restoration | Translocation from other sites, seeding, turfing, planting and cultivating. Compounds, use of plant and equipment, vehicular access, storage of materials, movement, soil and subsoil handling, testing and site investigations/surveys. |
| | After Care | Testing and site investigations/surveys, continuing effects of translocation from other sites, seeding, turfing, planting and cultivating. |
| | Ongoing Management | Restrictions on after use of land and ongoing management options as a result of project having occurred. |

★ Key information ★

**Box C.4.4**
**Suggested Criteria for Selecting Impact Assessment Methods**

**Sites Designated to Meet International Obligations**
Special Protection Areas (and pSPAs)
Special Areas of Conservation (and cSACs)
Sites of Community Importance
Ramsar Sites (and proposed Ramsar Sites)
World Heritage Sites
Biosphere Reserves
Biogenetic Reserves

**Nationally Designated Sites**
National Scenic Areas
Regional Parks
Historic Gardens and Designed Landscapes
National Nature Reserves

Sites of Special Scientific Interest
Geological Conservation Review Sites
Nature Conservation Review Sites
Marine Nature Reserves
Areas of Special Protection
EC Salmonoid and Cyprinid Fisheries
Aquifer Protection Zones
Environmentally Sensitive Areas (ESAs)

**Local Designations with a Statutory Basis**
Conservation Areas
Country Parks
Picnic Sites
Statutory Local Nature Reserves
Tree Preservation Orders

**Locally Designated Non-Statutory Designations**
Areas of Great/High Landscape Value
Other local landscape designations
Ancient Semi-Natural Woodland
Scottish Wildlife Trust Reserves
Woodland Trust Sites
Royal Society for the Protection of Birds Nature Reserves
Listed Wildlife Sites (SWT)
Sites of Importance for Nature Conservation (or local system variant)
Regionally Important Geological/ Geomorphological Sites

## C.5 Provision of Information

| Step in the EIA Process | |
|---|---|
| **Stage 1:** **Before Submission of the** **Environmental Statement** | Deciding whether EIA is required |
| | Requiring submission of an Environmental Statement |
| | Preliminary contacts and liaison |
| | Scoping the Environmental Statement |
| | **Information collection** |
| | Describing baseline environmental information |
| | Predicting environmental impacts |
| | Assessing the significance of impacts |
| | Mitigation measures and enhancement |
| | Presenting environmental information in the Environmental Statement |
| **Stage 2:** **Submission of Environmental** **Statement and Consideration of** **Environmental Information** | Submission of Environmental Statement and project application for consent |
| | Consultation and publicity |
| | Requiring more information |
| | Negotiating modifications to the project |
| | Considering the environmental information |
| **Stage 3:** **Making the Decision** | Making the decision |
| | Guaranteeing compliance |
| **Stage 4:** **Implementation** | Implementation of mitigation and compensation measures |
| | Monitoring |
| | Review, reassessment and remedial measures |
| | Reporting |

**[See also Figure 2, Sections C.1 and D.3, Appendices 1–6]**

**Table C.5.1 Summary of References for Equivalent Requirements in all EIA Regulations for Provision of Information by SNH**

| Topic | Regulations | Reference |
|---|---|---|
| Development requiring planning permission | EIASR 99 | Reg. 12 |
| Development by a PA including local roads | EIASR 99 | Regs 22–23 |
| Unauthorised development on appeal | EIASR 99 | Regs 32 |
| Review of old mineral permissions | EIASROMPRO2/EIASR99 | Regs 12 and 28A |
| Motorways and trunk roads | EIASR 99 | N/A |
| Drainage improvements | EIASR 99 | Reg. 58 |
| Marine aquaculture | EIAFishFarmMWR 99 | Reg. 7 |
| Forestry works | EIAForestrySR 99 | Reg. 12 |
| Use of uncultivated land and semi-natural areas for agriculture | ULSNARO2 | Reg. 8 |
| Irrigation, drainage and water management for agriculture | EIAWaterMR03/EIASR99 | Regs 12 |
| Electricity power stations >50MW and overhead lines | ElecWorks EIASR 00 | Regs 8 and 15 |
| Offshore electricity power stations >1MW | OffshoreGenStnsR02/ ElecWorks EIASR 00 | Regs 8 and 15 |
| Gas pipelines not requiring planning permission | PGasTransPWEIAR 99 | Reg. 9 |
| Offshore oil and gas and pipelines | OffshorePPPAEER 99 | Regs 5 and 8 |
| Other pipelines | PipelineWEIAR 00 | Reg. 6 |
| Decommissioning nuclear installations | NuclearREIADR 99 | Reg. 7 |
| Harbours, docks, piers and ferries | HarbourWEIAR 99 | N/A |

**C.5.1** The 'consultation bodies' (see Glossary in Annexe 1 below) have a statutory duty to take reasonable steps to organise and keep up to date any environmental information relevant to their functions, particularly that information listed in Reg. 4(2) of the Environmental Information (Scotland) Regulations 2004 (EISR04) and to actively and systematically disseminate the information to the public generally. They also have an explicit duty to provide relevant environmental information held by them to further the EIA process, particularly providing it to applicants and developers preparing an Environmental Statement (Reg. 5 EISR04 and Reg. 12 EIASR 99), but Reg. 12 of the EIASR 99 clearly states that any authority, body or person required to provide information under the EIASR 99 shall not be required to provide information which it is entitled or is bound to hold in confidence. They also have duties to provide advice and assistance to applicants as indicated below (C.5.9).

## Provision of Information

**C.5.2** In addition to the general duty of the consultation bodies to organise, keep up to date and disseminate environmental information (Reg. 4 EISR04) and to make environmental information available to all applicants (Regs 5 and 6 EISR04), the EIA Regulations also make provision for the mandatory release of environmental information, on request, to a developer (or their agents or consultants) preparing an Environmental Statement. It applies to all public bodies and specifically to statutory consultees (EIASR 99 Reg. 12) (see also Circular 15/1999, paragraphs 98–100).

**C.5.3** The duty to provide the information on request applies throughout the EIA process including the early stages of preparation of an Environmental Statement. A developer is not bound to provide the consultation bodies with full details of the project when asking for the information—it is sufficient to identify 'the land and the nature and purpose of the development' and the 'main environmental consequences to which the person giving the notice proposes to refer to in the Environmental Statement' (EIASR 99 Reg. 12).

**C.5.4** However, the EISR04 are more specific about the duties of the consultation bodies and what is expected of the applicant. Reg. 5(2) requires the public authority to provide information to an applicant on request in 20 working days and to ensure, as far as practicable, that the information provided is up to date, accurate and comparable (Reg. 5(3) EISR04). The applicant may request the information to be provided in a particular form or format (Reg. 6 EISR04) and the consultation bodies shall comply unless either it is reasonable to make the information available in another form or format, or it is publicly available and easily accessible to the applicant in another form or format.

**C.5.5** The 20 day period for supply of information may be extended by up to a further 20 days if the volume or complexity of the information requested makes it impractical for the consultation bodies to provide it in 20 days (Reg. 7 EISR04). The consultation bodies can make a reasonable charge for providing certain types of information (Reg. 8 EISR04).

**C.5.6** In exceptional cases, Reg. 10 makes provision for the consultation bodies to refuse to provide environmental information requested by an applicant, but these cases will be rare.

**C.5.7** The Competent Authorities have duties to inform the consultation bodies when they know of a case where the Regulations will apply but a developer may approach the consultation bodies before the Competent Authority, and they have a duty to provide the information requested, if the developer says it is in connection with the EIA Regulations.

**C.5.8**  Developers should not simply ask for all information held by a consultation body for a particular site or area. The developer may consult the consultation bodies to see whether they hold information relevant to the Environmental Statement. The Regulations require the consultation bodies to 'enter into consultation with that person to determine whether it has in its possession any information which he or they consider relevant to the preparation of the Environmental Statement and, if they have, the public authority shall make that information available to that person' (EIASR 99 Reg. 12(4)).

### Duty to Provide Advice and Assistance

**C.5.9**  Reg. 9 of the EISR04 requires the consultation bodies to provide advice and assistance, so far as reasonable, to applicants and prospective applicants. Where a request for information has been formulated in too general a manner, the consultation bodies shall ask the applicant as soon as possible, and in any event within the 20 day period, to provide more particulars in relation to the request and should assist the applicant in providing those particulars. However, if the consultation bodies operate in accordance with a code of practice produced by the Scottish Ministers under Reg. 18 EISR04, the duty to provide advice and assistance will be deemed to have been met by compliance with the code.

**C.5.10**  The EIA Regulations do not override the EISR04, but sit alongside them and are intended to be complementary to them. Both Regulations seek to apply the requirements of EC Directives (in the case of the Environmental Information Regulations, via *The Freedom of Information Scotland Act* 2002 which itself is intended to comply with the EC Directive 2003/4/EC on Public Access to Environmental Information) (29).

### Projects Using Uncultivated Land and Semi-Natural Areas for Intensive Agriculture (ULSNA)

**C.5.11**  Reg. 8 provides that, if consulted on a scoping opinion or requested by the applicant, the consultation bodies must determine whether they have information relevant to the preparation of the Environmental Statement and, if so, must make it available within 28 days of the request, unless it is capable of being, or required to be, treated as confidential. The consultation bodies may make a reasonable charge for providing the information. Where a consultation body was consulted on a scoping opinion it must advise the applicant that it holds relevant information and the cost of making it available but need only make it available if requested by the applicant.

## C.6 Describing Baseline Environmental Information

| | Step in the EIA Process |
|---|---|
| **Stage 1:**<br>**Before Submission of the**<br>**Environmental Statement** | Deciding whether EIA is required<br>Requiring submission of an Environmental Statement<br>Preliminary contacts and liaison<br>Scoping the Environmental Statement<br>Information collection<br>**Describing baseline environmental information**<br>Predicting environmental impacts<br>Assessing the significance of impacts<br>Mitigation measures and enhancement<br>Presenting environmental information in the environmental statement |
| **Stage 2:**<br>**Submission of Environmental**<br>**Statement and Consideration of**<br>**Environmental Information** | Submission of Environmental Statement and project application for consent<br>Consultation and publicity<br>Requiring more information<br>Negotiating modifications to the project<br>Considering the environmental information |
| **Stage 3:**<br>**Making the Decision** | Making the decision<br>Guaranteeing compliance |
| **Stage 4:**<br>**Implementation** | Implementation of mitigation and compensation measures<br>Monitoring<br>Review, reassessment and remedial measures<br>Reporting |

**[See also Figure 2, Section B.4, Appendices 1–6]**

### Statutory Provisions and Government Guidance

**C.6.1**   Contributing to the analysis of baseline information is a non-statutory part of the process. However, the developer must include the information in the Environmental Statement so this is a necessary procedure for the developer. Guidance on this stage is also provided in PAN 58 at paragraphs 38–39 and 44–46.

### The Developer's Responsibilities

**C.6.2**   Collecting baseline information on the environment ought to be a relatively straightforward part of the EIA process (compared to impact prediction and other aspects) but it is often done inadequately. Unless there is a clear understanding of the baseline and how that may change without the changes that would be brought about by the project, there is little hope of the Environmental Statement accurately predicting and mitigating the impacts of the development.

**C.6.3**   Information gathering should be comprehensive in respect of the significant environmental issues to be addressed in the Environmental Statement. Field work should be carefully planned, bearing in mind the seasonal constraints on some work such as ornithological, botanical, landscape and archaeological surveys. Environmental information sources should be identified and the relevant central and local government authorities and agencies should be consulted. Local communities and voluntary bodies should also be consulted as these groups can provide invaluable information.

**C.6.4**    Appendices 1–6 of this Handbook set out the best practice guidelines based on published work. This section sets out:

a.  common problems and pitfalls (Box C.6.1);

b.  good EIA practice (Box C.6.2); and

c.  advice on ensuring an integrated approach to the natural heritage.

★ **Good EIA practice** ★

**Box C.6.1**

**Baseline Information: Good EIA Practice Avoids these Common Problems and Pitfalls**

- Reliance on existing recorded data only.

- Insufficient time to conduct surveys at appropriate seasons/times.

- Inadequate expertise in surveys.

- Lack of understanding of what information is needed to inform the EIA process.

- Inadequate resources for baseline surveys leading to incomplete or inept results.

- Use of out of date material.

- Lack of verification of collated information.

- Omission of important information that is available/obtainable.

- Lack of an adequate national/regional context e.g. of Landscape Character Types.

- Too narrow a focus on the site, paying insufficient attention to landscape, natural features, processes or influences of surrounding land.

- Use of inappropriate techniques or inappropriate application of appropriate methods of survey e.g. landscape character assessment, NVC, Phase 1 Habitat Surveys etc.

- Concentration on the easier aspects of survey e.g. birds and mammals, whilst ignoring difficult ones such as invertebrates or bryophytes which may be better indicators of environmental conditions.

- Inadequate acknowledgement of data limitations.

- Omission, lack of understanding or misrepresentation of designations, their purpose, reasons for designations and implications.

**C.6.5**    Wherever ecological impacts are expected to affect botanical interests or habitats supporting animal species of interest, vascular plants should normally be surveyed to at least establish NVC communities as this information is likely to be needed to inform ecological assessment. In habitats where lower plants are important constituents of the vegetation (for example moorland, Sphagnum mires) bryophytes and lichens should also be surveyed. For similar reasons, benthic communities should be included in marine surveys. Landscape character assessments are an essential prerequisite to effective landscape impact assessment.

## Box C.6.2

### Baseline Information: Good EIA Practice

- Thorough scoping of baseline data requirements and available information.

- Consultation and agreement on survey subjects, methods and emphasis.

- Use of best available information.

- Identification of influences on baseline information that would lead to change in absence of the project.

- Recruitment to the EIA team, temporarily if necessary, of people with skills and experience of field surveys in all relevant fields.

- Correct timing of surveys with adequate timescales to record variations in differing circumstances.

- Careful verification and validation of existing records with an appropriate balance between use of documentary and field survey material.

- Inclusion of likely changes that would be brought about by other projects already consented but not yet implemented.

- Consideration of baseline information which would contribute to assessment of cumulative, offsite, indirect impacts etc.

- Clear identification and agreement as to the appropriate level of detail of surveys and information gathering.

- Relating all baseline studies to their relevance to the nature, size, duration and location of the project to ensure all relevant information is collated without submerging it in a volume of irrelevant or over-detailed information.

- Early recognition of gaps in information and limitations in data that can be collated and consideration of how these gaps and limitations will be dealt with in the Environmental Statement.

## Box C.6.3

### Field Surveys

The developer should undertake field surveys in every case where natural heritage effects are likely to be significant or effects cannot be predicted at the scoping stage. Where relevant, landscape and visual surveys, habitat and species surveys, surveys of natural features and processes, and outdoor recreation/access surveys will be essential to adequately inform landscape, visual, ecological, earth heritage and outdoor recreation impacts in Environmental Statements.

**C.6.6**    Where a long lead time on the Environmental Statement allows, it may be possible to monitor changes in existing conditions prior to the submission of the Environmental Statement. This would allow trends in ecological or landscape change or natural processes to be investigated and should be encouraged, although it is rarely possible to do this.

### Integrating Natural Heritage Issues

**C.6.7**    Owing to the different professional skills involved, it is common practice in Environmental Statements to address natural heritage issues separately, for example:

Landscape and Visual Impacts

Ecological Impacts

Impacts on the Marine Environment, Marine Systems and Coastal Processes

Cultural Heritage: Historic Gardens and Designed Landscapes

Geological and Soil Impacts: Earth Heritage

Public Amenity/Recreation Impacts: Outdoor access.

**C.6.8**    In many Environmental Statements even these sections or chapters can be subdivided, each being written by a separate author with specialist knowledge of, for example, aquatic or terrestrial ecology. In order to ensure authoritative assessment the practice of different authors each presenting their conclusions should be encouraged, but the Environmental Statement team co-ordinator should ensure that all of these differing elements are consistent and drawn together in an integrated and understandable presentation.

---

★ **Key advice** ★

**Box C.6.4**

**The Approach to Baseline Information**

When discussing or commenting on a (draft) Environmental Statement, Competent Authorities and consultees should encourage rigorous assessment by appropriately qualified and experienced professionals with specialists used where appropriate, and the facility in the Environmental Statement for all of their respective assessments to be clearly and consistently set out.

However, Competent Authorities and consultees should also encourage an integrated approach to natural heritage issues. The inter-relationships between landscape, visual, ecological and earth heritage information and the implications for the enjoyment of, access to and better appreciation of the natural heritage should be clearly set out.

Competent Authorities and consultees should encourage different aspects of the natural heritage to be assessed on a common basis wherever possible. For example landscape and ecological assessments may be able to use the same broad scales of significance so the significance of the different effects on the natural heritage can be directly compared.

# C.7 Predicting Environmental Impacts

| | Step in the EIA Process |
|---|---|
| **Stage 1:** **Before Submission of the** **Environmental Statement** | Deciding whether EIA is required |
| | Requiring submission of an Environmental Statement |
| | Preliminary contacts and liaison |
| | Scoping the Environmental Statement |
| | Information collection |
| | Describing baseline environmental information |
| | **Predicting environmental impacts** |
| | Assessing the significance of impacts |
| | Mitigation measures and enhancement |
| | Presenting environmental information in the Environmental Statement |
| **Stage 2:** **Submission of Environmental** **Statement and Consideration of** **Environmental Information** | Submission of Environmental Statement and project application for consent |
| | Consultation and publicity |
| | Requiring more information |
| | Negotiating modifications to the project |
| | Considering the environmental information |
| **Stage 3:** **Making the Decision** | Making the decision |
| | Guaranteeing compliance |
| **Stage 4:** **Implementation** | Implementation of mitigation and compensation measures |
| | Monitoring |
| | Review, reassessment and remedial measures |
| | Reporting |

**[See also Figure 2, Sections C.3, C.4, Appendices 1–6]**

## Statutory Provisions and Government Guidance

**C.7.1**    A prediction of environmental effects must be included in an Environmental Statement so this is a necessary procedure for the developer. Guidance on this stage is also provided in PAN 58 at paragraphs 47–52.

## Impact Prediction

**C.7.2**    Predicting and describing significant environmental impacts is a statutory requirement to include in an Environmental Statement. Reference is made to C.4 above and Appendices 1–6 of this Handbook.

**C.7.3**    Predicting the effects of a proposed project is a fundamental stage in EIA. One of the main purposes of the Environmental Statement is to clearly explain what the impacts of a proposal would be. The impacts should always be included in the non-technical summary in a way that is understandable to the general public. However, this is not always easy in respect of natural heritage implications.

**C.7.4**    Predicting environmental impacts involves 2 main elements of work:

- Anticipating, modeling, predicting or forecasting the changes that would be brought about by the project at all of its life stages.

- Explaining, in a rational, consistent, impartial and transparent way, the significance of the changes.

**C.7.5**    Changes or effects are usually referred to as 'impacts'.

**C.7.6**    The effectiveness of impact prediction in Environmental Statements varies considerably. Given the constraints of sometimes inadequate available information, the evolving nature of modelling and predictive techniques, the lack of understanding as to how the environment may respond to some impacts and the extensive reliance of the process on professional judgement, it is not surprising that this element of the EIA process has been widely criticised. Research (25), (26), (27) shows a more rigorous and more impartial assessment of predicted effects in many Environmental Statements since 1992. The trend is one of improvement but some Environmental Statements are still weak in this area.

---

★ **Good EIA practice** ★

**Box C.7.1**

**The Approach to Impact Prediction**

Competent Authorities and consultees should adopt a practical and rational approach to commenting on the effectiveness of impact prediction. If they are unable to support the findings, criticism should be focused on key issues rather than detail. As a minimum Competent Authorities and consultees should try to ensure that an Environmental Statement fairly and consistently describes

a.  the sensitivity of the environmental resource;
b.  the magnitude of change in absolute terms where possible and relative terms elsewhere;
c.  the likelihood of the impacts occurring;
d.  the certainty with which impacts have been identified;
e.  the comparison with the do nothing alternative (see C.8.4 below) and other alternative solutions that are feasible and practical;
f.  the significance of the impacts based on the factors (a)–(d) above.

---

**C.7.7**    Appendices 1–6 of this Handbook contain more detailed advice on best practice techniques for predicting impacts and assessing and explaining their significance. It is likely that Competent Authorities will need specialist advice in respect of some aspects of EIA, from the consultation bodies and others.

---

★ **Good EIA practice** ★

**Box C.7.2    Types of Impact**

The effects of a proposal may be:

| | |
|---|---|
| predictable or unpredictable; | certain or uncertain; |
| direct or indirect; | avoidable or unavoidable; |
| positive (beneficial) or negative (harmful); | reversible or irreversible; |
| | localised or widespread; |
| temporary or permanent; | small or large; |
| short, medium or long-term; | individual or cumulative; |
| one-off, intermittent or continuous; | significant or of no consequence. |
| immediate or delayed; | |

**C.7.8**    Different effects may be experienced at different stages in a project's life (e.g. site preparation, construction, operation, decommissioning or restoration (see also Figure 3)). The Environmental Statement should clearly set out the effects on the natural heritage and their interrelationships with each other and with other environmental effects.

**C.7.9**    This will usually require factual information. Prediction of impacts should be as objective and, where possible, as quantified as possible. However, there will often be uncertainties so a range of potential results may need to be considered with an explanation about the nature of the uncertainties associated with predictions.

**C.7.10**    The Information required for impact prediction will generally include:

- the likelihood of the impact occurring at the magnitude anticipated;

- the likely duration of the impact and whether it would be continuous, intermittent, immediate or delayed;

- the extent to which the impact could be reversed;

- the feasibility and effectiveness of any measures designed to mitigate the impact;

- the cumulative effects of different impacts in this project;

- the cumulative effects of the same impact in this and other projects;

- the cumulative effects of different impacts in this and other projects; and

- the risk and effects of unscheduled, emergency or accidental events and the effects of the resulting activity.

**C.7.11**    The magnitude of change should generally be expressed in absolute terms and relatively in terms of percentage change to habitat area or species population or net gains and losses of important landscape features. Given the likelihood of uncertainties, the degree of confidence in the predictions as to the magnitude of effects should also be indicated. The status of the site will generally be a factual expression of the international, national, regional or local importance of landscape, habitats or species. The sensitivity of the landscape, habitats and species will require a professional and sometimes subjective judgement, usually taking account, for example, of the distribution, population, rarity or vulnerability to change of the habitats and species in nature conservation terms and the vulnerability of landscapes to loss of local character or distinctiveness.

**C.7.12**    By way of example, Figure 4 is an illustration of a matrix showing the magnitude of changes in the landscape. Landscape impact magnitude is based, amongst other things, on the extent of change to the landscape resource, the duration, scale and nature of the change and the impact of the change on the character of the landscape and its tolerance for accommodating change. This is an example only, each EIA will require its own matrix designed to meet the particular circumstances.

**C.7.13** The impacts should be considered in the light of any information available or reasonably obtainable about the capacity of environments to accommodate change. Limits of acceptable change can sometimes be defined and these are particularly relevant to EIA procedures.

---

**Figure 4**

**Example of Scale of Magnitude of Changes to the Landscape Resource**

| | |
|---|---|
| High magnitude | Significant changes, over a significant area, to key characteristics or features or to the landscape's character or distinctiveness for more than 2 years |
| Medium magnitude | Noticeable but not significant changes for more than 2 years or significant changes for more than 6 months but less than 2 years, over a significant area, to key characteristics or features or to the landscape's character or distinctiveness. |
| Low magnitude | Noticeable changes for less than 2 years, significant changes for less than 6 months, or barely discernible changes for any length of time. |
| No change | No predicted changes. |

---

**C.7.14** Where limits cannot or should not be defined, a broader approach, assessing the capacity of habitats or landscapes to accommodate change, in more general, relative terms could be used. The SNH national programme of Landscape Character Assessments is a particularly important resource contributing to the EIA process. These assessments should be used in every case. They are the best available baseline information for landscape assessment and the most authoritative source of comment on the sensitivity of landscapes, based more on their character and distinctiveness. Assessment should focus on landscape character rather than designations, although designations will need to be considered in the light of their policy implications (see Section C.6 above).

**C.7.15** Natural Heritage Resource Assessments would also provide authoritative and comprehensive source information relating to the natural heritage resource in an integrated way. These too should be used in EIA to help provide a sound context for the site assessment (SNH, 1996, *Assessing the Natural Heritage Resource Guidance Note for Local Authorities*).

## C.8 Assessing the Significance of Impacts

| Step in the EIA Process | |
|---|---|
| **Stage 1:** **Before Submission of the Environmental Statement** | Deciding whether EIA is required |
| | Requiring submission of an Environmental Statement |
| | Preliminary contacts and liaison |
| | Scoping the Environmental Statement |
| | Information collection |
| | Describing baseline environmental information |
| | Predicting environmental impacts |
| | **Assessing the significance of impacts** |
| | Mitigation measures and enhancement |
| | Presenting environmental information in the Environmental Statement |
| **Stage 2:** **Submission of Environmental Statement and Consideration of Environmental Information** | Submission of Environmental Statement and project application for consent |
| | Consultation and publicity |
| | Requiring more information |
| | Negotiating modifications to the project |
| | Considering the environmental information |
| **Stage 3:** **Making the Decision** | Making the decision |
| | Guaranteeing compliance |
| **Stage 4:** **Implementation** | Implementation of mitigation and compensation measures |
| | Monitoring |
| | Review, reassessment and remedial measures |
| | Reporting |

**[See also Figure 2, Sections C.3, C.4 and C.7 and Appendices 1–6]**

### Statutory Provisions and Government Guidance

**C.8.1** The Environmental Statement should include a description of the nature, scale and significance of the effects, so this is a necessary procedure for the developer. It will also be a necessary procedure for consultees to consider the significance of the effects in order to make representations to the Competent Authority. Guidance on this stage is also provided in PAN 58 at paragraphs 47–52.

### The Significance of Impacts

**C.8.2** Whereas prediction of impacts should be a largely objective step, assessing the significance of impacts relies, at least in part, on value judgements, including placing weight or value on the environment likely to experience the change. The significance of impacts at this stage should relate back to the impacts deemed to be significant at the scoping stage (section C.4 above). It is also possible that new environmental effects may come to light in the assessment process because it should be iterative. Essentially, the EIA was undertaken to address impacts that were then deemed to be significant; has it revealed that the impacts will occur and if so how important will they be?

**C.8.3** The significance of change is also related to the duration, timing and extent of effects, the degree of certainty in the prediction of impacts and the likelihood of irreversible changes occurring. For example, an effect which is unlikely, or the likelihood of which is uncertain, may nevertheless be significant if it would be a very serious or irreversible adverse effect if it did occur. This is the basis of the 'precautionary principle', see Section D.1 below.

**C.8.4**    The significance of the effects of a proposed development should be considered in the context of changes that will occur regardless of whether the project goes ahead or not, the 'do-nothing' alternative. The 'do-nothing' comparison, or in some cases, such as road improvements, the 'do-minimum' comparison,  is a projection of the existing data to provide a baseline for comparison to show how the site would change if the project did not go ahead. The 'do-nothing' comparison examines trends currently occurring at the site, including likely management, land use changes or other interventions, and assesses the significance of these changing conditions. The 'do-nothing' comparison, however, should be used in a reasonable way, genuinely predicting likely change and not taking the best possible comparison for the purpose of the Environmental Statement.

**C.8.5**    Alternative solutions, if the project went ahead in a different form or at a different location, should normally be considered. This will reveal the full picture of the project's effects and the least damaging option. If alternatives have been considered they should be included in the Environmental Statement with an explanation why they were rejected.

★ **Key information** ★

**Box C.8.1**

**Factors Affecting Significance of Impacts**

The significance of an impact is derived from an analysis of:

● the sensitivity of the environment to change, including its capacity to accommodate the kinds of changes the project may bring about;

● the amount and type of change, often referred to as the impact magnitude which includes the timing, scale, size and duration of the impact;

● the likelihood of the impact occurring—which may range from certainty to a remote possibility;

● comparing the impacts on the environment which would result from the project with the changes that would occur without the project—often referred to as the 'do-nothing' comparison; and

● Expressing the significance of the impacts of the project, usually in relative terms, based on the principle that the more sensitive the resource, the more likely the changes and the greater the magnitude of the changes, compared with the do-nothing comparison, the greater will be the significance of the impact.

**C.8.6**    A matrix can be used for considering the significance of impacts. This may combine the work previously undertaken for the assessment in respect of baseline information about the resource and impact prediction. The sensitivity of the resource can be analysed from the baseline information and may be summarised and classified in a matrix, an example of which is given in Figure 5 below.

**C.8.7**    The significance matrix can combine the information about the sensitivity of the resource, in this case the landscape resource, with the information previously compiled about the magnitude of impacts, of the kind shown in Figure 4 above. Combining the 2 sets of analysis, from Figures 4 and 5, enables a simple matrix of significance to be compiled as shown in Figure 6.

---

**Figure 5**
## Example of Scale of Sensitivity of Landscape Receptors

| | |
|---|---|
| High Sensitivity | Key characteristics and features, identified by systematic landscape character assessment, which contribute significantly to the distinctiveness and character of the landscape character type. Designated landscapes e.g. National Parks, NSAs and AGLVs and landscapes identified as having low capacity to accommodate proposed form of change. |
| Medium Sensitivity | Other characteristics or features of the landscape that contribute to the character of the landscape locally. Locally valued landscapes which are not designated. Landscapes identified as having some tolerance of the proposed change subject to design and mitigation etc. |
| Low Sensitivity | Landscape characteristics and features that do not make a significant contribution to landscape character or distinctiveness locally, or which are untypical or uncharacteristic of the landscape type. Landscapes identified as being generally tolerant of the proposed change subject to design and mitigation etc. |

---

**Figure 6**
## Example of a Matrix Showing Impact Significance Related to Sensitivity and Magnitude of Change

| Significance of Impact | Sensitivity of Receptor | Magnitude of Change |
|---|---|---|
| Substantial/High | High | High |
| | High | Medium |
| | Medium | High |
| Moderate/Medium | High | Low |
| | Medium | Medium |
| | Low | High |
| Slight/Low | Medium | Low |
| | Low | Medium |
| | Low | Low |
| No Change | High, Medium or Low | No Change |

**C.8.8**  The construction of the matrix for weighing the significance of landscape and visual impacts should be adapted to fit individual cases or types of cases. For example, a significance matrix for natural heritage impacts may look like the example in Figure 6. The impacts are individually ranked for their significance on the basis of the sensitivity of the resource and the magnitude of the change, a high sensitivity resource and high magnitude of change would result, self evidently, in a high or 'substantial' significance of the impact.

**C.8.9**  Beneficial and adverse impacts should be treated in the same way.

★ Key advice ★

**Box C.8.2**

**The Approach to Impact Significance**

Competent Authorities and consultees should ensure that all Environmental Statements:

● clearly set out the sensitivity of the natural heritage resource;

● clearly set out the magnitude and likelihood of change, compared with at least the baseline information but preferably compared with the do-nothing alternative; and

● explain the significance of all relevant impacts on the natural heritage in a systematic, impartial, consistent and rational way that is clearly described in the Environmental Statement.

**C.8.10**  Predicting impact significance is partly objective and partly subjective. It relies on the professional judgement of landscape architects, ecologists and others who may place varying weight on the many factors involved. This naturally leads to differences of opinion. The Environmental Statement should therefore set out the basis of these judgements so that others can see the weight attached to different factors and can understand the rationale of the assessment. The Environmental Statement should clearly explain how the impact significance has been derived.

★ Key advice ★

**Box C.8.3**

**Comments on Significance**

Consultees should not seek to criticise an Environmental Statement merely because it expresses conclusions which do not accord with their conclusions.

Wherever possible, comments should identify why the conclusions are different so the Competent Authority may judge the basis of the 2 different assessments.

Consultees should indicate how and where impact prediction in the Environmental Statement is inappropriate, e.g. where:

● inappropriate predictive techniques have been used;
● impacts have been omitted;
● the sensitivity of the resource is under-estimated (e.g. insufficient attention has been paid to reasons why areas have been designated);
● any aspect of the timing, scale, size or duration of the impact has been omitted or inappropriately applied to the assessment;
● the impacts are not compared adequately or appropriately with the do-nothing or other relevant alternatives;
● the scale of impact significance is unclear, inconsistent, inappropriate or partial.

## C.9  Mitigation Measures and Enhancement

| Step in the EIA Process | |
|---|---|
| **Stage 1:**<br>**Before Submission of the**<br>**Environmental Statement** | Deciding whether EIA is required<br>Requiring submission of an Environmental Statement<br>Preliminary contacts and liaison<br>Scoping the Environmental Statement<br>Information collection<br>Describing baseline environmental information<br>Predicting environmental impacts<br>Assessing the significance of impacts<br>**Mitigation measures and enhancement**<br>Presenting environmental information in the Environmental Statement |
| **Stage 2:**<br>**Submission of Environmental**<br>**Statement and Consideration of**<br>**Environmental Information** | Submission of Environmental Statement and project application for consent<br>Consultation and publicity<br>Requiring more information<br>Negotiating modifications to the project<br>Considering the environmental information |
| **Stage 3:**<br>**Making the Decision** | Making the decision<br>Guaranteeing compliance |
| **Stage 4:**<br>**Implementation** | Implementation of mitigation and compensation measures<br>Monitoring<br>Review, reassessment and remedial measures<br>Reporting |

**[See also Figure 2, Sections B.6, C.4, D.7, D.10 and E.2 and Appendices 1–6]**

### Statutory Provisions and Government Guidance

**C.9.1**    The Environmental Statement must include a description of the mitigation measures, so this is a necessary procedure for the developer. It will also be a necessary procedure for consultees to consider the effectiveness of mitigation in order to make representations to the Competent Authority.

**C.9.2**    Mitigation measures are a statutory requirement to include in an Environmental Statement, enhancement is not. Reference is made to B.6, C.4 above and Appendices 1–6 of this Handbook. Guidance on this stage is also provided in PAN 58 at paragraphs 53–61.

### Introduction

**C.9.3**    One of the main purposes of EIA is to ensure that potentially significant environmental effects of proposed projects are avoided or reduced as far as possible or practicable. This can be achieved by many different measures which might include:

● locating the project so as not to affect environmentally sensitive locations;

● using construction, operation and restoration methods or processes which reduce environmental effects;

- designing the whole project carefully to avoid or minimise environmental impacts;

- introducing specific measures into the project design, construction, decommissioning and restoration that will reduce or compensate for adverse effects.

**C.9.4**    In the EIA process it will be necessary to consider four distinct treatments of the project and its environmental effects as follows:

      Avoidance
      Reduction
      Remedy/Compensation
      Enhancement/Net Benefit.

**C.9.5**    These distinctions are not merely of academic interest. Avoidance, reduction, and remedy/compensation are all mitigation measures in the context of the EIA Regulations (they have different meanings in the context of the *Conservation (Natural Habitats &c) Regulations* 1994). They must, therefore, be included and described in every Environmental Statement (see sections B.6 and C.4 above). Enhancement, or net benefit, or new benefit, may be offered by the developer. Often an Environmental Statement has claimed enhancement but the measures are not genuine enhancement because the loss or damage to the natural heritage is greater than the benefit of the 'enhancement' proposed or the measures are more akin to compensatory measures.

**C.9.6**    In any event, competent authorities need to distinguish between mitigating measures and enhancement to clearly understand the effectiveness of the mitigation.

> ★ **Key information** ★
>
> ### Box C.9.1
>
> ### Mitigation and Enhancement
>
> Mitigating measures or mitigation are the measures taken to avoid, reduce or remedy adverse impacts of the project.
>
> Avoidance is the measures taken to avoid any adverse impacts, including alternative or 'do-nothing' options.
>
> Reduction is the measures taken to reduce unavoidable adverse impacts of the project.
>
> Remedial or compensatory measures or compensation are other measures taken to (at least try to) offset or compensate for residual adverse effects which cannot be avoided or further reduced.
>
> Enhancement/net benefit/new benefit is the genuine enhancement of the environmental interest of a site or area because adverse effects are limited in scope and scale, and the project includes improved management or new habitats or features, which are better than the prospective management, or the habitats or features present there now. There is, therefore, a net or new benefit to the natural heritage.

**C.9.7**    The distinction is also relevant to consultees. For example, a project may result in adverse ecological effects on existing habitats, that cannot be mitigated, e.g. loss of an important peatland area, but may also result in genuine landscape enhancement elsewhere. The one is not a compensation for the other. The loss and the benefit must be weighed as separate issues. Likewise, a Competent Authority may need to weigh the significance of harm to the natural heritage perhaps with enhancement of other environmental conditions.

**C.9.8**    Developers are entitled to include environmental enhancement in their Environmental Statement. Whilst most Environmental Statements tend to focus on mitigation, developers may perceive an opportunity to help to persuade a Competent Authority to grant consent by offering some form of enhancement, to tip the balance in favour of the project.

**C.9.9**    In many cases there will be opportunities to encourage enhancement of the natural heritage, especially where the existing ecological interest is low or a landscape has been degraded. Mineral or waste restoration schemes and woodland planting schemes often offer potential for genuine enhancement where the harm to the natural heritage is insignificant.

---

★ Key advice ★

**Box C.9.2    The Approach to Mitigation**

Generally, Competent Authorities and consultees should promote a sustainable approach and give priority to:

- firstly avoiding adverse impacts on the natural heritage; then

- reducing unavoidable adverse effects on the natural heritage; then

- compensating for the adverse effects that cannot be further reduced; and

- in parallel with this prioritised approach to mitigation, encouraging opportunities to enhance the natural heritage.

---

**C.9.10**    However, it should be borne in mind that enhancement cannot be insisted upon.

**C.9.11**    The effectiveness of mitigating measures, their reliability and certainty, and the commitment to ensuring their practical implementation should be addressed in the Environmental Statement (See Section D.3). The environmental effects of mitigating measures themselves should also be assessed. Measures may have been added at a late stage and may not have been assessed in the Environmental Statement. The measures themselves may have significant environmental effects, for example through further habitat loss or by the obstruction of wildlife corridors or intrusion into the landscape or obstruction of views.

**C.9.12**    The effectiveness of measures such as habitat recreation, restoration, revegetation or habitat or species translocation should be considered on their merits in the circumstances of each case. However, bearing in mind the general experience of habitat and species translocation, this should normally be regarded as a last resort when destruction of individuals of the species is inevitable, that is, a rescue operation.

## C.10  Presenting Environmental Information in the Environmental Statement

| | Step in the EIA Process |
|---|---|
| **Stage 1:**<br>**Before Submission of the**<br>**Environmental Statement** | Deciding whether EIA is required<br>Requiring submission of an Environmental Statement<br>Preliminary contacts and liaison<br>Scoping the Environmental Statement<br>Information collection<br>Describing baseline environmental information<br>Predicting environmental impacts<br>Assessing the significance of impacts<br>Mitigation measures and enhancement |
| | **Presenting environmental information in the Environmental Statement** |
| **Stage 2:**<br>**Submission of Environmental**<br>**Statement and Consideration of**<br>**Environmental Information** | Submission of Environmental Statement and project application for consent<br>Consultation and publicity<br>Requiring more information<br>Negotiating modifications to the project<br>Considering the environmental information |
| **Stage 3:**<br>**Making the Decision** | Making the decision<br>Guaranteeing compliance |
| **Stage 4:**<br>**Implementation** | Implementation of mitigation and compensation measures<br>Monitoring<br>Review, reassessment and remedial measures<br>Reporting |

**[See also Figure 2 and Section B.6 above and Appendices 1–6 of this Handbook]**

### Statutory Provisions and Government Guidance

**C.10.1**   There are statutory duties on developers to include certain matters in an Environmental Statement (see B.6 above). There are statutory powers for Competent Authorities to require the inclusion of certain matters in an Environmental Statement. However, the way in which these matters are presented in the Environmental Statement is a matter for the developer, a non-statutory procedure, which may involve discussion with consultees. The Regulations do not specify how environmental information should be presented in an Environmental Statement, except to say that a non-technical summary must be included. In practice, non-technical summaries are often separately bound and more widely distributed and available. Guidance on this stage is also provided in PAN 58 at paragraphs 72–79.

### Presentation

**C.10.2**   Presentation therefore depends largely on the importance of the various issues in the Environmental Statement. Where no significant natural heritage issues arise the Environmental Statement may simply refer to them in a general chapter on other environmental effects or information. Where natural heritage issues are significant they should be addressed to the extent necessary in the main body of the Environmental Statement, although larger Environmental Statements may have separate volumes containing detailed information about specific issues. Topic reports in Appendices are a common and accepted feature of Environmental Statements.

**C.10.3**   The size of an Environmental Statement will depend on the range and complexity of issues and no standard size can be given. However, the Institute of Environmental Management and Assessment consider that Environmental Statements with less than 50 pages are usually regarded as inadequately detailed if more than one or 2 key topics are involved. Environmental Statements of more than 150 pages should only be necessary where the project has many environmental impacts and is of a large scale. Too much detail can distract and deter readers and make key issues difficult to appreciate.

★ **Good EIA practice** ★

**Box C.10.1**

**Good EIA Practice**

**Presentation of Environmental Statements**

Environmental Statements should be:

● adequate for the purpose but succinct and not over-detailed;

● clear and understandable;

● consistent in content and presentation across issues;

● well, but not lavishly, presented with the effective use of maps, diagrams, charts, drawings, illustrations, photographs, sketches, photo montages, tables and matrices to reduce text and explain complex issues and with summaries and key conclusions highlighted;

● scientifically sound, but with the minimum use of scientific and technical language, with glossaries and the use of common names for species and an annexe for scientific nomenclature wherever possible;

● inclusive of source data to allow readers to interpret this for themselves but with detailed information in appendices or separate volumes to avoid cluttering the main text of the assessment;

● logical in its structure, presenting a clear description of the project, baseline information, prediction of effects and their significance, before mitigation measures, and then describing the mitigation measures and the residual effects of the project (including their significance) taking mitigation into account;

● free standing and not reliant on key information in another document especially if that document is not publicly available;

● based wherever possible on standard methods or standard forms of presentation that will be familiar at least to other specialists or professionals advising the Competent Authority.

**C.10.4**   Environmental Statements are increasingly available on CD or DVD and distribution in this form is compliant subject to the caveats explained in paragraphs D.1.8 and D.1.9 below.

# Part D

# Consideration of the Environmental Statement

(and the Project Consent Application)

## D.1 Submission of the Environmental Statement and Project Application for Consent: the Roles of the Competent Authority, the Developer and Consultees

| | Step in the EIA Process |
|---|---|
| **Stage 1:**<br>**Before Submission of the**<br>**Environmental Statement** | Deciding whether EIA is required<br>Requiring submission of an Environmental Statement<br>Preliminary contacts and liaison<br>Scoping the Environmental Statement<br>Information collection<br>Describing baseline environmental information<br>Predicting environmental impacts<br>Assessing the significance of impacts<br>Mitigation measures and enhancement<br>Presenting environmental linformation in the Environmental Statement |
| **Stage 2:**<br>**Submission of Environmental**<br>**Statement and Consideration of**<br>**Environmental Information** | **Submission of Environmental Statement and project application for consent**<br>Consultation and publicity<br>Requiring more information<br>Negotiating modifications to the project<br>Considering the environmental information |
| **Stage 3:**<br>**Making the Decision** | Making the decision<br>Guaranteeing compliance |
| **Stage 4:**<br>**Implementation** | Implementation of mitigation and compensation measures<br>Monitoring<br>Review, reassessment and remedial measures<br>Reporting |

**[See also Sections B.2 and E.2, Attachment A and Annexe 2]**

### Statutory Provisions and Government Guidance

**D.1.1**   With the exceptions relating to harbours, docks, piers and jetties, the Competent Authority has a statutory duty to consult the consultation bodies and to publicise every Environmental Statement. The consultation bodies should respond in every case. The form of their response is not prescribed in the Regulations, which refer only to them making 'representations'.

### The Competent Authority's Role

**D.1.2**   The Competent Authority is the body responsible for making the decision whether the project should be given a consent, permission, licence or other authorisation. It may be the Scottish Ministers, a planning authority, SEPA or other statutory authorities such as the Forestry Commission.

**D.1.3**   With regard to their duties in respect of an Environmental Statement a Competent Authority **must:**

a. register and publicise the application and Environmental Statement as required by the Regulations and take account of any representations received from the public;

b. notify other bodies and consult in accordance with the Regulations and take account of any representations received;

c. not make a decision on the application for at least 4 weeks; and

d. not grant consent or other authorisation unless they have taken account of the environmental information;

e. if granting consent, record on the face of the permission or other authorisation that they have taken account of the environmental information;

f. notify their decision in accordance with the Regulations.

**D.1.4** The Competent Authority is responsible for evaluating the Environmental Statement to ensure it addresses all of the relevant environmental issues and that the information is presented accurately, clearly and systematically. The Competent Authority should be prepared to challenge the findings of the Environmental Statement if it believes they are not adequately supported by scientific evidence. If it believes that key issues are not fully addressed, or not addressed at all, it **must** request further information. The authority has to ensure that it has in its possession **all** relevant environmental information about the likely significant environmental effects of the project **before** it makes its decision whether to grant permission. It is too late to address the issues after permission has been granted.

**D.1.5** The Competent Authority **may** also:

a. seek and take the advice or representations of bodies other than the statutory consultees;

b. require the developer to submit further environmental information;

c. refuse the application;

d. grant consent subject to conditions or limitations over and above those set out in the Environmental Statement and the application.

**D.1.6** The developer must submit sufficient copies of the Environmental Statement to the Competent Authority to enable them to undertake the statutory consultations and, in addition, 3 copies for the Scottish Ministers, one of which will be deposited in the Scottish Executive library where a full collection of all Environmental Statements submitted in Scotland is available to the public.

**D.1.7** Under the provisions of Article 21 of and Sch. 11 to the Town and Country Planning (Electronic Communications) (Scotland) Order 2004 (TCPECSO04), environmental statements may be distributed electronically and (with the exceptions noted at D.1.8 below) notices under the EIASR99 will be deemed to have been given on condition that the electronic communication (e mail and attachment(s)) of the document (statement or notice):

a. is capable of being accessed by the recipient; and

b. is legible in all material respects, that is, it is as readable as if it were available to the recipient in hard copy (see further definition at regulation 2A(5) of the EIASR 99 added by the TCPECSO04); and

c. is sufficiently permanent that it can be used for subsequent reference.

**D.1.8**   Electronic communication cannot be used in the EIA process in respect of a developer serving any notice under regulation 13 on those with an interest in neighbouring land, or in respect of any transboundary consultation with other EC Member States, or in respect of any unauthorised development which is going through the EIA process retrospectively.

**D.1.9**   Environmental Statements are increasingly available on CD or DVD and distribution in this form is compliant subject to the above caveats.

## Consultees and the public need to be clear about the development applied for

**D.1.10**   In all cases, it is important that  it is clear as to what the development is that is applied for. In granting consent, the Competent Authority will permit the development applied for as described in the application and the plans submitted with it (subject to any conditions or modifications); this may or may not be exactly the same as the development described and assessed in the Environmental Statement. It may be important to differentiate between information in the Environmental Statement about the proposal (the planning application) and information on the environmental impacts of the proposal (the EIA). Where there is any discrepancy between information on the application plans and information in the Environmental Statement, it is the information in the plan that will normally prevail and which will be granted permission.

## D.2 Consultation and Publicity

| | Step in the EIA Process |
|---|---|
| **Stage 1:** **Before Submission of the Environmental Statement** | Deciding whether EIA is required |
| | Requiring submission of an Environmental Statement |
| | Preliminary contacts and liaison |
| | Scoping the Environmental Statement |
| | Information collection |
| | Describing baseline environmental information |
| | Predicting environmental impacts |
| | Assessing the significance of impacts |
| | Mitigation measures and enhancement |
| | Presenting environmental information in the Environmental Statement |
| **Stage 2:** **Submission of Environmental Statement and Consideration of Environmental Information** | Submission of Environmental Statement and project application for consent |
| | **Consultation and publicity** |
| | Requiring more information |
| | Negotiating modifications to the project |
| | Considering the environmental information |
| **Stage 3:** **Making the Decision** | Making the decision |
| | Guaranteeing compliance |
| **Stage 4:** **Implementation** | Implementation of mitigation and compensation measures |
| | Monitoring |
| | Review, reassessment and remedial measures |
| | Reporting |

**D.2.1**    The Competent Authority has a statutory duty to consult the 'consultation bodies', and to publicise every Environmental Statement.

**D.2.2**    The Regulations prescribe the procedures to be adopted by Competent Authorities in respect of consultations and publicity. Every Environmental Statement must be accessible to the public and must be publicised. Planning related Environmental Statements must be placed on deposit in the planning authority's office for at least 4 weeks, and must be advertised by notices in newspapers to enable the public to make representations about the project and its environmental effects and to comment on the Environmental Statement. Neighbouring landowners, occupiers and lessees must also be notified (EIASR 99 Regs 13–17; also Circular 15/1999 paragraphs 101–112. Guidance on this stage is also provided in PAN 58 at paragraphs 24–27).

**D.2.3**    In addition, to ensure compliance with the Directive, the Regulations require some Competent Authorities to consult certain bodies in respect of every Environmental Statement and other bodies in respect of particular types of Environmental Statement or where a project is in a particular type of location.

### The Statutory Consultees

**D.2.4**    The statutory consultees (where they are not the Competent Authority making the decision) include:
The Scottish Ministers
The Planning Authority
Adjacent planning authorities whose area may be affected
The Scottish Environmental Protection Agency (SEPA
Scottish Natural Heritage

| Topic | Regulations | Ref to consult | Time period |
|---|---|---|---|
| Development requiring planning permission | EIASR 99 | Regs 14 + 16 | 4 weeks |
| Development by a PA including local roads | EIASR 99 | Reg. 24 | Unspecified |
| Unauthorised development on appeal | EIASR 99 | Reg. 35 | Unspecified |
| Review of old mineral permissions | EIASROMPRO2/EIASR 99 | Regs 14, 16 and 28A | 4 weeks |
| Motorways and trunk roads | EIASR 99 | Regs 49 + 50 and S.20A and 55A Roads (Scotland) Act 1980 | Opportunity to express an opinion |
| Drainage improvements | EIASR 99 | Reg. 59 | 28 days |
| Marine aquaculture | EIAFishFarmMWR 99 | Reg. 9 | 28 days |
| Forestry works | EIAForestrySR 99 | Regs 20–23 | 28 days |
| Use of uncultivated land and semi-natural areas for agriculture | ULSNARO2 | Reg. 9(2) | 42 days |
| Irrigation, drainage and water management for agriculture | EIAWaterMRO3/EIASR99 | Regs 14 + 16 | 4 weeks |
| Electricity power stations >50MW and overhead lines | ElecWorks EIASR 00 | Reg. 11 | 14 days from receipt of Env. Statement |
| Offshore electricity power stations >1MW | OffshoreGenStnsRO2/ ElecWorks EIASR 00 | Reg. 11 | 14 days from receipt of Env. Statement |
| Gas pipelines not requiring planning permission | PGasTransPWEIAR 99 | Reg. 10 | 28 days |
| Offshore oil and gas and pipelines | OffshorePPPAEER 99 | Regs 9 + 10 | 4 weeks |
| Other pipelines | PipelineWEIAR 00 | Reg. 7 | 28 days |
| Decommissioning nuclear installations | NuclearREIADR 99 | Regs 8–9 | Such reasonable time as HSE may specify |
| Harbours, docks, piers and ferries | HarbourWEIAR 99 | Regs 7 + 9 and Sch. 3 (14–15) Harbours Act 1964 | Reasonable opportunity |

**[See also Sections C.4, D.1 and D.8 and Annexe 2]**
**Statutory Provisions and Government Guidance**

The Health and Safety Executive but not for roads EIA
Historic Scotland
the water and sewerage authority or authorities for the area but not for roads EIA.

**D.2.5**    These statutory consultees have a duty to provide the developer, on request, with any relevant information in their possession, which may assist in the preparation of the Environmental Statement, EIASR 99 Reg. 12 (see also Section C.5 above).

**D.2.6**    The Regulations also require that consultees are informed of the submission of an Environmental Statement in conjunction with a development application, supplied with a copy of the Environmental Statement and given an opportunity to comment on its contents. Such comments should be supplied to the Competent Authority to assist in the decision. The time allowed to respond is generally 28 days (4 weeks) from the date or notice (which is considered here to be the date of receipt by them as there can be no notice until the consultation has been received). Exceptions to this are summarised in Table D.2.1 below.

**D.2.7**    There is no statutory provision for consultation with members of the general public during the preparation of an Environmental Statement. However, if the developer informs the Competent Authority that an Environmental Statement will be prepared, then the fact that EIA is under way must be published. Developers may, however, legally prepare an Environmental Statement without informing the Competent Authority or statutory consultees beforehand. If this occurs, the consultees must be informed upon the Authority's receipt of the Environmental Statement. There are3obligations on developers in this regard.

a.  A notice should be placed in a local newspaper by the planning authority advertising the deposit of the Environmental Statement and its availability and the developer must pay the cost of this publicity.

b.  A reasonable number of copies of the Environmental Statement should be made available to the public (EIASR 99 Reg. 17) but a charge may be made (EIASR 99 Reg. 18).

c.  Notice must be given to everyone with a legal interest in neighbouring land.

**D.2.8**    Electronic communication cannot be used for the notices under D.2.9(c) above, but the statement can be distributed electronically in accordance with the TCPECSR04 (see further D.1.8–9 above).

**D.2.9**    It should be noted that there is no specific provision dealing with amendments or additions to an Environmental Statement that has already been submitted. Where an applicant changes an Environmental Statement, rather than simply providing further information, which is very specifically defined in the Regulations, the safest approach is to treat any addition or amendment as an Environmental Statement submitted during the course of a planning application and to advertise the whole of the Environmental Statement, with the amendment/addition, in compliance with regulation 13 EIASR99. This will ensure compliance with the general intent of the EIA Directive to notify and inform people of the possible environmental effects of a proposed development.

## D.3 Liaison with the Competent Authority and the Developer

| Step in the EIA Process | |
|---|---|
| **Stage 1:** **Before Submission of the Environmental Statement** | Deciding whether EIA is required |
| | Requiring submission of an Environmental Statement |
| | Preliminary contacts and liaison |
| | Scoping the Environmental Statement |
| | Information collection |
| | Describing baseline environmental information |
| | Predicting environmental impacts |
| | Assessing the significance of impacts |
| | Mitigation measures and enhancement |
| | Presenting environmental information in the Environmental Statement |
| **Stage 2:** **Submission of Environmental Statement and Consideration of Environmental Information** | Submission of Environmental Statement and poject application for consent |
| | **Consultation and publicity** |
| | Requiring more information |
| | Negotiating modifications to the project |
| | Considering the environmental information |
| **Stage 3:** **Making the Decision** | Making the decision |
| | Guaranteeing compliance |
| **Stage 4:** **Implementation** | Implementation of mitigation and compensation measures |
| | Monitoring |
| | Review, reassessment and remedial measures |
| | Reporting |

**[See also Sections C.4, D.1 and D.8 and Annexe 2]**

### Statutory Provisions and Government Guidance

**D.3.1**   The Competent Authority has a statutory duty to consult the consultation bodies, and to publicise every Environmental Statement.

**D.3.2**   Consultees should maintain liaison with the Competent Authority as may be necessary in the circumstances of each case. Sometimes it will be sufficient to respond to the application and Environmental Statement in writing in one step. Often, however, there will be advantages in a dialogue between consultees and the Competent Authority and often the developer too. EIA, at its best, is an interactive process with each of the main parties informing and influencing the others.

★ **Key advice** ★

**Box D.3.1  Liaison**

If a consultee seeks more information, for example, this should be through the Competent Authority but, in exceptional circumstances, for example where that Authority is slow or reluctant to request the information, the consultee may approach the developer directly; however, in such cases it is vital that the information is submitted to the Competent Authority, not just the consultee.

Dialogue and liaison between consultees, the Competent Authority and the developer will generally improve understanding of the project, the environmental issues, the effects of the project and the views of the consultees. It will usually increase the effectiveness of the EIA process and the influence of consultees on the decision.

Correspondence between the developer and consultees should normally be copied to the Competent Authority.

## D.4   Wider Consultation and Dissemination

| | Step in the EIA Process |
|---|---|
| **Stage 1:** <br> **Before Submission of the** <br> **Environmental Statement** | Deciding whether EIA is required <br> Requiring submission of an Environmental Statement <br> Preliminary contacts and liaison <br> Scoping the Environmental Statement <br> Information collection <br> Describing baseline environmental information <br> Predicting environmental impacts <br> Assessing the significance of impacts <br> Mitigation measures and enhancement <br> Presenting environmental information in the Environmental Statement |
| **Stage 2:** <br> **Submission of Environmental** <br> **Statement and Consideration of** <br> **Environmental Information** | Submission of Environmental Statement and project application for consent <br> **Consultation and publicity** <br> Requiring more information <br> Negotiating modifications to the project <br> Considering the environmental information |
| **Stage 3:** <br> **Making the Decision** | Making the decision <br> Guaranteeing compliance |
| **Stage 4:** <br> **Implementation** | Implementation of mitigation and compensation measures <br> Monitoring <br> Review, reassessment and remedial measures <br> Reporting |

## [See also Sections B.1, D.1, D.2, and D.3 and Annexe 2]

### Statutory Provisions and Government Guidance

**D.4.1**    The Competent Authority has a statutory duty to consult the consultation bodies, and to publicise every Environmental Statement. Guidance on this stage is also provided in PAN 58 at paragraphs 26–27.

### Wider Consultation

**D.4.2**    It is a matter for the Competent Authority to decide who should be consulted beyond the statutory consultees. However, it is open to consultees to suggest or recommend that other bodies should also be notified or consulted. This is particularly important where other bodies are known to hold important and relevant information and/or expertise.

**D.4.3**    It may be convenient to share copies of the application and Environmental Statement. Copying Environmental Statements may be restricted by copyright; do not copy without the permission of the authors or developer. Many consultants or developers will supply further copies, either free or at a reasonable cost, or other bodies could go to the locations where the Statement is lodged. Environmental Statements are increasingly available on CD or DVD and distribution in this form is compliant subject to the caveats in D.1.8 above.

**D.4.4**    Even where a consultee has involved other bodies their comments should be submitted separately. Even if the Competent Authority declined to consult directly, the other bodies, nevertheless, have a right to submit representations to the Competent Authority direct. Their representations must be taken into account, as environmental information in the meaning of the Regulations.

# D.5  Transboundary Environmental Effects

| Step in the EIA Process | |
|---|---|
| **Stage 1:** **Before Submission of the Environmental Statement** | Deciding whether EIA is required |
| | Requiring submission of an Environmental Statement |
| | Preliminary contacts and liaison |
| | Scoping the Environmental Statement |
| | Information collection |
| | Describing baseline environmental information |
| | Predicting environmental impacts |
| | Assessing the significance of impacts |
| | Mitigation measures and enhancement |
| | Presenting environmental information in the Environmental Statement |
| **Stage 2:** **Submission of Environmental Statement and Consideration of Environmental Information** | Submission of Environmental Statement and project application for consent |
| | **Consultation and publicity** |
| | Requiring more information |
| | Negotiating modifications to the project |
| | Considering the environmental information |
| **Stage 3:** **Making the Decision** | Making the decision |
| | Guaranteeing compliance |
| **Stage 4:** **Implementation** | Implementation of mitigation and compensation measures |
| | Monitoring |
| | Review, reassessment and remedial measures |
| | Reporting |

## Table D.5.1  Summary of References for Requirements in all EIA Regulations for Transboundary Effects

| Topic | Regulations | Reference |
|---|---|---|
| Development requiring planning permission | EIASR 99 | Regs 40–41 |
| Development by a PA including local roads | EIASR 99 | Regs 40–41 |
| Unauthorised development on appeal | EIASR 99 | Regs 39–41 |
| Review of old mineral permissions | EIASROMPRO2/EIASR99 | Regs 40–41 and 28A |
| Motorways and trunk roads | EIASR 99 | Reg. 49 |
| Drainage improvements | EIASR 99 | Reg. N/A |
| Marine aquaculture | EIAFishFarmMWR 99 | Reg. N/A |
| Forestry works | EIAForestrySR 99 | Reg. 14 |
| Use of uncultivated land and semi-natural areas for agriculture | ULSNARO2 | Reg. 12 |
| Irrigation, drainage and water management for agriculture | EIAWaterMRO3/EIASR99 | Regs 40–41 |
| Electricity power stations >50MW and overhead lines | ElecWorks EIASR 00 | Reg. 12 |
| Offshore electricity power stations >1MW | OffshoreGenStnsRO2/ ElecWorks EIASR 00 | Reg. 12 |
| Gas pipelines not requiring planning permission | PGasTransPWEIAR 99 | Reg. 13 |
| Offshore oil and gas and pipelines | OffshorePPPAEER 99 | Regs 5 + 12 |
| Other pipelines | PipelineWEIAR 00 | Reg. 3 |
| Decommissioning nuclear installations | NuclearREIADR 99 | Regs 8 + 12 |
| Harbours, docks, piers and ferries | HarbourWEIAR 99 | Reg. 8 |

**[See also Sections D.1, D.2, D.3 and D.4 and Annexe 2]**

## Statutory Provisions and Government Guidance

**D.5.1**    Guidance on the Government's procedures for transboundary EIA are given in paragraphs 119 to 121 of Circular 15/1999 and statutory requirements are in Regs 40–41 EIASR 99.

**D.5.2**    Electronic communication cannot be used for transboundary consultations see further D.1.8 above.

## Projects Using Uncultivated Land and Semi-Natural Areas for Intensive Agriculture (ULSNA)

**D.5.3**    The usual kinds of provisions for consultations on transboundary effects are contained in Reg. 11. However, recognising that projects could span the border, Reg. 12 makes explicit provision for transborder cases. Essentially, whichever country the greater part of the application site lies in will determine the regulations to be applied. Thus, if more of the site lies in Scotland, the Scottish regulations will apply. However, there are provisions in Reg. 12(3 B 5) for agreement amongst the respective governments and consultation bodies for the procedures of either the English or Scottish regulations to apply even where the lesser part of the site lies in that jurisdiction.

## D.6    Requiring More Information or Analysis

| Step in the EIA Process | |
| --- | --- |
| **Stage 1:**<br>**Before Submission of the**<br>**Environmental Statement** | Deciding whether EIA is required<br>Requiring submission of an Environmental Statement<br>Preliminary contacts and laison<br>Scoping the Environmental Statement<br>Information collection<br>Describing baseline environmental information<br>Predicting environmental impacts<br>Assessing the significance of impacts<br>Mitigation measures and enhancement<br>Presenting environmental information in the Environmental Statement |
| **Stage 2:**<br>**Submission of Environmental**<br>**Statement and Consideration of**<br>**Environmental Information** | Submission of Environmental Statement and project application for consent<br>Consultation and publicity<br>**Requiring more information**<br>Negotiating modifications to the project<br>Considering the environmental information |
| **Stage 3:**<br>**Making the Decision** | Making the decision<br>Guaranteeing compliance |
| **Stage 4:**<br>**Implementation** | Implementation of mitigation and compensation measures<br>Monitoring<br>Review, reassessment and remedial measures<br>Reporting |

**[See also Figure 2, Sections C.3, C.4, D.7 and D.8, Appendices 1–6, Attachment A and Annexe 2]**

**D.6.1**    The Competent Authority has the statutory power to require certain additional information to be submitted by the developer. (See Reg. 19 EIASR 99 and Circular 15/1999, paragraphs 113–118. Guidance on this stage is also provided in PAN 58 at paragraphs 91–92).

### Requiring Information

**D.6.2**    If important information, which could affect the outcome of the application, is absent or inadequate consultees should inform the Competent Authority as soon as possible. They should ask the Competent Authority to require the applicant to submit the information, if necessary as a supplementary Environmental Statement (see D.8 below), and ask the authority not to determine the application until all of the necessary environmental information is available. Submission of the required information may mean that you have to reassess the natural heritage impacts of the proposal.

**D.6.3**    It is important to obtain any further information via the Competent Authority; but, in exceptional circumstances, for example where that Authority is slow or reluctant to request the information, consultees may approach the developer directly. However, in such cases it is vital that the information is submitted to the Competent Authority, not just the consultee.

**D.6.4**    A consultee's response is required primarily on whether the project should be consented or authorised and, if so, on what terms and conditions and, if not, why not. A consultee should not risk its views being too late to influence the

**Table D.6.1  Summary of References for Requirements in all EIA Regulations for Transboundary Effects**

| Topic | Regulations | Reference |
|---|---|---|
| Development requiring planning permission | EIASR 99 | Reg. 19 |
| Development by a PA including local roads | EIASR 99 | Reg. 24 |
| Unauthorised development on appeal | EIASR 99 | Reg. 36 |
| Review of old mineral permissions | EIASROMPRO2/EIASR99 | Regs 19 and 28A |
| Motorways and trunk roads | EIASR 99 | N/A |
| Drainage improvements | EIASR 99 | Reg. 60 |
| Marine aquaculture | EIAFishFarmMWR 99 | Reg. 10 |
| Forestry works | EIAForestrySR 99 | Reg. 11 |
| Use of uncultivated land and semi-natural areas for agriculture | ULSNARO2 | Regs 9 + 10 |
| Irrigation, drainage and water management for agriculture | EIAWaterMRO3/EIASR99 | Reg. 19 |
| Electricity power stations >50MW and overhead lines | ElecWorks EIASR 00 | Reg. 13 |
| Offshore electricity power stations >1MW | OffshoreGenStnsRO2/ ElecWorks EIASR 00 | Reg. 13 |
| Electricity power stations and overhead lines | ElecWorks EIASR 00 | Reg. 13 |
| Gas pipelines not requiring planning permission | PGasTransPWEIAR 99 | Reg. 11 |
| Offshore oil and gas and pipelines | OffshorePPPAEER 99 | Reg. 10 |
| Other pipelines | PipelineWEIAR 00 | Reg. 8 |
| Decommissioning nuclear installations | NuclearREIADR 99 | Reg. 10 |
| Harbours, docks, piers and ferries | HarbourWEIAR 99 | N/A |

decision merely because they are awaiting a response to a request for more information.

**D.6.5**    It is important to bear in mind that the costs and delays involved in obtaining and submitting additional information can be considerable. EIA is not an opportunity to obtain information that is desirable for other purposes, although information obtained as a necessity in an EIA case may, of course, contribute to environmental knowledge generally.

★ Key advice ★

**Box D.6.1**
**Requests for Further Information**

Information should only be requested when it is essential, not merely desirable, to the decision on the project and where it could actually influence a consultee's or a Competent Authority's views in a substantive way.

Requests for additional information should have regard to the feasibility and practicality of obtaining it and the timescale and cost.

Requests should be reasonable.

They should normally be made via the Competent Authority.

They should be made promptly and in one step if possible.

**D.6.6** Repeated requests for more and more information can be very frustrating to the developer and Competent Authority and can indicate a lack of clarity of thought initially on the part of the body that keeps requesting. However, it is reasonable to expect that, from time to time, new information may prompt an obvious need for yet further information.

## Requesting Further Information for Full Planning Applications

**D.6.7** In respect of full planning applications, the planning authority has the power to require information to be submitted under 2 statutory provisions, namely:

a. Regulation 19 of the EIASR 99, requiring submission of:

i)  any further environmental information to enable the application to be determined; or

ii)  information concerning any matter which is required to be dealt with in the Environmental Statement (i.e. matters in Schedule 4 EIASR 99); or

iii)  information reasonably required to give proper consideration to the application; or

iv)  evidence to verify any information in the Environmental Statement;

b.  Article 13 of the Town and Country Planning (General Development Procedure) (Scotland) Order 1992 requiring any further information in order to enable them to deal with the application.

**D.6.8** When requesting planning authorities to require further information, reference can be made to the EIASR 99, the General Development Procedure (Scotland) Order 1992 (GDPO), para 48 of Circular 15/1999, and the letter from Scottish Executive Development Department to all Heads of Planning in Scotland, dated June 2002.

**D.6.9** A request for further information on the planning application should be made under Articles 13 and 4(3) of the GDPO. A request for further work on the EIA should be made under the EIASR 99. Thus, a clear distinction should be made wherever possible, **further information about the proposal that forms the subject of the planning application should be obtained by means of the GDPO; further information on the environmental impacts of the proposal should be obtained by means of the EIASR 99.**

**D.6.10** If the applicant refuses to make available further information and/or the planning authority will not support a consultee's request, or the planning authority agrees with the applicant that further information is not necessary, the consultee will have to accept that the information will not be obtained. The result may be that they have no choice but to object to the application and, if necessary, ask that the case be referred to the Scottish Ministers for their own determination.

**D.6.11** A consultee should always take care to identify what further information is required and should be able to justify this request clearly. Most planning authorities will use their powers to require further information to be submitted rather than merely going straight for a refusal of permission.

## Requesting Further Information for Outline Planning Applications

**D.6.12**  The EIA process, in relation to planning, applies only at the stage of the grant of planning permission, either full or outline planning permission. The subsequent approval of reserved matters following outline planning permission is not a grant of permission, so the EIA process is not applicable. A condition cannot be imposed on an outline planning permission requiring the submission of an Environmental Statement at approval of reserved matters stage. The environmental information must be considered before the grant of outline planning permission. It follows that there should be an adequate Environmental Statement at the outline application stage. Where an outline application is submitted without an Environmental Statement or with an inadequate Environmental Statement the matter must be remedied before the outline permission is granted. (See Circular 15/1999, paragraph 48.)

**D.6.13**  Where outline planning permission is sought, it may not be possible to predict impacts on the natural heritage with the information submitted, because the details of the development are unknown. Therefore, the information in the EIA can only inform the decision in principle on whether development of the nature proposed is acceptable at all on the site. If the information available in the Environmental Statement at this stage is insufficient to determine whether the development is acceptable in principle, the planning authority should require such information to be submitted as it is reasonably necessary to assess the likely environmental effects of the proposal or they should refuse planning permission, possibly with an indication that a detailed application would be considered if it is supported by an Environmental Statement. (See also paragraphs C.1.29–31 and section C.4 above and para 48 of Circular 15/1999.)

**D.6.14**  In respect of outline planning applications, the planning authority has the power to require information to be submitted under the same provisions as described in D.6.9–11 above, plus the power to require reserved matters to be submitted under the provisions of Article 4(3) of the Town and Country Planning (General Development Procedure) (Scotland) Order 1992. This enables a planning authority to require the submission of all or any reserved matters that the planning authority considers it to be necessary to consider before the grant of an outline planning permission. However, whereas there is no limit to the period in which the other provisions can be used to require information to be submitted, there is a 1 month time limit on the use of Article 4(3) requiring reserved matters to be submitted.

**D.6.15**  **Thus, requests for some of the reserved matters to be submitted must be made by the planning authority within 1 month under Article 4(3) of the GDPO; requests for further information about the proposal that forms the subject of the planning application should be obtained (at any time) by means of the GDPO; and further information on the environmental impacts of the proposal should be obtained (at any time) by means of Regulation 13 of the EIASR 99.**

## Projects Using Uncultivated Land and Semi-Natural Areas for Intensive Agriculture (ULSNA)

**D.6.16**  Reg. 10 ULSNAR02 provides for SEERAD to require further information to be submitted where, after complying with the requirements in Reg. 9(2) to consult and publicise the Environmental Statement, it determines that the

Environmental Statement must include additional information in order for it to be an Environmental Statement. The applicant must provide the additional information. The consultation bodies will be consulted on the additional information and generally have 28 days in which to comment.

# D.7 Requiring Modifications of the Project

| Step in the EIA Process | |
| --- | --- |
| **Stage 1:** **Before Submission of the Environmental Statement** | Deciding whether EIA is required |
| | Requiring submission of an Environmental Statement |
| | Preliminary contacts and liaison |
| | Scoping the Environmental Statement |
| | Information collection |
| | Describing baseline environmental information |
| | Predicting environmental impacts |
| | Assessing the significance of impacts |
| | Mitigation measures and enhancement |
| | Presenting environmental information in the Environmental Statement |
| **Stage 2:** **Submission of Environmental Statement and Consideration of Environmental Information** | Submission of Environmental Statement and project application for consent |
| | Consultation and publicity |
| | Requiring more information |
| | **Negotiating modifications to the project** |
| | Considering the environmental information |
| **Stage 3:** **Making the Decision** | Making the decision |
| | Guaranteeing compliance |
| **Stage 4:** **Implementation** | Implementation of mitigation and compensation measures |
| | Monitoring |
| | Review, reassessment and remedial measures |
| | Reporting |

**[See also Sections C.9, D.6 and D.8, E.3 and E.4, Attachment A and Annexe 2]**

## Modifications

**D.7.1**    The fact that a project is subject to the EIA process does not preclude modifications or amendments. These may be made during the EIA process, after the Environmental Statement has been submitted. Indeed, the iterative process of EIA is very likely to lead to consultees or the Competent Authority seeking changes to the proposals to avoid or further reduce environmental effects. Equally, the proposer may wish to make changes to help satisfy concerns expressed by the Competent Authority, consultees or the public.

**D.7.2**    Where they would meet a consultee's concerns, modifications should be encouraged both before and after the consultation response has been submitted (indeed, it may be the consultation responses that initiate the discussions about modifications).

**D.7.3**    It is open to a consultee, throughout the EIA process to negotiate modifications to a project via the Competent Authority. If necessary, the consultee can ask a developer to modify the project if the Competent Authority will not require the developer to do so (see D.6.5 above). However, if the developer does agree to modify the project, it is vital that the modification is submitted formally to the Competent Authority. Modifications provided solely to the consultee, and not to the Planning Authority as an amendment to the application will not constitute any part of the planning application, nor will they constitute any part of the planning permission.

**Box D.7.1**
**Negotiating Changes**

A consultee should consider whether to open negotiations, with the Competent Authority and/or the developer, to affect changes to the proposals, if they conclude that:

a. more or different mitigation would be appropriate;
b. adverse effects could be avoided, or compensated; or
c. new benefits could be achieved.

**D.7.4**   Technically speaking, there are no procedures by which an application, e.g. a planning application, can be modified after it has been submitted. Theoretically, the application should be withdrawn and resubmitted. However, for fairly obvious reasons, most Competent Authorities take a practical approach. They accept modifications and ensure that it is clear which scheme is consented, if a consent is issued (e.g. by imposing a condition referring to revised plans). From a procedural point of view, whether the modifications can be accepted as an amendment, without a new application being made, is a decision for the Competent Authority alone.

**D.7.5**   The key questions will be:

a.  whether the modifications are so extensive as to amount to a different project proposal–in which case a new application should be made; or

b.  whether the modifications are significant but not extensive–in which case a new application is generally not required but the Competent Authority should reconsult and renotify and republicise the proposal; or

c.  whether the modifications are not so significant as to merit reconsultation and republicising generally, but may be appropriate for selected consultees to comment, or whether no consultees need comment.

**D.7.6**   It should be noted that there is no specific provision dealing with amendments or additions to an Environmental Statement that has already been submitted. Where an applicant changes a proposal, and therefore has to change the Environmental Statement, rather than simply providing further information (which is very specifically defined in the Regulations) any addition or amendment should be treated as an Environmental Statement submitted during the course of a planning application. The Competent Authority should advertise the whole of the Environmental Statement, with the amendment/addition, in compliance with regulation 13 EIASR99. This will ensure compliance with the general intent of the EIA Directive to notify and inform people of the possible environmental effects of a proposed development.

## D.8  Is a Supplementary Environmental Statement Required?

| | Step in the EIA Process |
|---|---|
| **Stage 1:**<br>**Before Submission of the**<br>**Environmental Statement** | Deciding whether EIA is required<br>Requiring submission of an Environmental Statement<br>Preliminary contacts and liaison<br>Scoping the Environmental Statement<br>Information collection<br>Describing baseline environmental information<br>Predicting environmental impacts<br>Assessing the significance of impacts<br>Mitigation measures and enhancement<br>Presenting environmental information in the Environmental Statement |
| **Stage 2:**<br>**Submission of Environmental**<br>**Statement and Consideration of**<br>**Environmental Information** | Submission of Environmental Statement and project application for consent<br>Consultation and publicity<br>**Requiring more information**<br>**Negotiating modifications to the project**<br>Considering the environmental information |
| **Stage 3:**<br>**Making the Decision** | Making the decision<br>Guaranteeing compliance |
| **Stage 4:**<br>**Implementation** | Implementation of mitigation and compensation measures<br>Monitoring<br>Review, reassessment and remedial measures<br>Reporting |

**[See also Sections B.4, D.1, D.6, D.7 and E.4, Attachment A and Annexe 2]**

### Statutory Provisions and Government Guidance

**D.8.1**    A Competent Authority has the statutory powers to require further environmental information to be submitted. This may be in the form of a revised or supplementary Environmental Statement or otherwise. However, it should be noted that submission of documents called supplementary or revised Environmental Statements is a convention, which often occurs in practice, but not a statutory process or term. The EIASR 99 only refer to 'further information' being required and submitted and treats this more as a part of the overall environmental information than as a new Environmental Statement. Such further information should also be submitted in respect of significant modifications to a project, but see further D.7 above.

**D.8.2**    Where the Competent Authority has decided to accept a modification it will need to consider whether a new or revised or supplementary Environmental Statement is necessary. Essentially the question to be asked will be 'is the project still the project that was assessed in the original Environmental Statement or a different project in ways that mean the effects of the project have not been adequately assessed?' The legal cases referred to in Annexe 9 will be relevant here. Guidance on this stage is also provided in PAN 58 at paragraphs 26–27.

**D.8.3**    The key point is that the final decision on the application must take account of the environmental information for that project, as it would be consented.

It follows that it would not be appropriate to consider environmental information about another form of the project.

## Supplementary Environmental Statement

**D.8.4** A supplementary Environmental Statement is submitted where the original Environmental Statement was incomplete or further work on environmental effects has been undertaken (whether or not the project has been modified since the original application and Environmental Statement were submitted). A supplementary Environmental Statement may be submitted, to add to the original, to ensure that all of the relevant environmental information is considered by the Competent Authority. The supplementary Environmental Statement may include a revision of the whole or part of the original document or additions that are needed to cover the additional information.

**D.8.5** Where the additional information is submitted because the Competent Authority required it to be, the supplementary Environmental Statement must follow the same procedures as for the original Statement, including publicity and consultations (Regulation 19 EIASR 99 and Circular 15/1999 paragraphs 114–117). However, where the supplementary Statement or other environmental information is submitted voluntarily the Competent Authority has discretion as to the extent of publicity and consultation it undertakes.

## Revised Environmental Statement

**D.8.6** A revised Environmental Statement is submitted where a project has been modified since the original application and Environmental Statement were submitted. A revised Environmental Statement may be submitted to amend the original, to ensure that the environmental information considered by the Competent Authority relates to the project as modified. The revised Environmental Statement may be a revision of the whole of the original document or a revision of those parts of the original Environmental Statement that need to be changed as a result of the modifications.

**D.8.7** Again, where the additional information is submitted because the Competent Authority required it to be, the revised Environmental Statement must follow the same procedures as for the original Statement, including publicity and consultations (Regulation 19 EIASR 99 and Circular 15/1999 paragraphs 114–117). Where the revised Statement or other environmental information is submitted voluntarily the Competent Authority has discretion as to the extent of publicity and consultation it undertakes.

## Deciding about Submissions

**D.8.8** Deciding the extent to which environmental information should be resubmitted as a result of modifications to the project is sometimes difficult to ascertain. There are no statutory provisions for procedures and a Competent Authority may need help from the consultees in deciding whether:

a. the project is so extensively different that a new application and new Environmental Statement is required; or

b. the project is significantly different and the Environmental Statement should be revised (with consultation on the revision) or added to by a supplementary Environmental Statement (with consultation following); or

c. the project and its environmental effects are not so significantly different as to invalidate the original Environmental Statement and consultation and publicity responses as being the environmental information relevant and appropriate to the project at the point in time of the decision.

**D.8.9**    The following points may be of assistance in these decisions:

a. submission of an entirely new area of environmental information should generally be in a supplementary Environmental Statement;

b. submission of revisions relating to changes to the environmental information contained in the Environmental Statement should be submitted either as a revised Environmental Statement (in whole or in part) replacing the original or in a supplementary Environmental Statement (if additional information is included) adding to the original Environmental Statement and in whole or in part replacing it;

c. minor revisions which correct, update or otherwise amend information in the original Environmental Statement but do not change any of the impacts in terms of their significance could be submitted as an Addendum or Erratum as appropriate.

**D.8.10**    Generally, there should be no need to reconsult or renotify Addenda or Errata unless a particular consultee has requested the changes and/or the changes may be significant to one or more particular interests.

**D.8.11**    Like the decision on whether to require an Environmental Statement in the first instance, the decision whether a new or supplementary or revised Environmental Statement is required, and the procedures for dealing with the submission, consultation, publicity etc., are all a matter for the Competent Authority.

★ **Key advice** ★

**Box D.8.1**
**Involvement of Consultees in Procedures for Dealing with Revised or Supplementary Environmental Statements**

Consultees may urge a Competent Authority to adopt a particular requirement or procedure in respect of revised or supplementary Environmental Statements, but cannot force them to. How a Competent Authority deals with revisions or supplementary information is a matter for the Competent Authority.

However, if important matters are at stake, and a consultee believes a serious breach of the Regulations is likely to result, for example by actually denying the public or statutory and other agencies the opportunity of commenting on changes to an Environmental Statement, they should refer the matter to the Scottish Ministers.

# D.9  Reviewing the Environmental Statement

| | Step in the EIA Process |
|---|---|
| **Stage 1:**<br>**Before Submission of the**<br>**Environmental Statement** | Deciding whether EIA is required<br>Requiring submission of an Environmental Statement<br>Preliminary contacts and liaison<br>Scoping the Environmental Statement<br>Information collection<br>Describing baseline environmental information<br>Predicting environmental impacts<br>Assessing the significance of impacts<br>Mitigation measures and enhancement<br>Presenting environmental information in the Environmental Statement |
| **Stage 2:**<br>**Submission of Environmental**<br>**Statement and Consideration of**<br>**Environmental Information** | Submission of Environmental Statement and project application for consent<br>Consultation and publicity<br>Requiring more information<br>Negotiating modifications to the project<br>Considering the environmental information |
| **Stage 3:**<br>**Making the Decision** | Making the decision<br>Guaranteeing compliance |
| **Stage 4:**<br>**Implementation** | Implementation of mitigation and compensation measures<br>Monitoring<br>Review, reassessment and remedial measures<br>Reporting |

**[See also Sections D.10, E.1, E.2, E.3 and E.4, Appendices 1–6, Attachment A and Annexe 2]**

## Statutory Provisions and Government Guidance

**D.9.1**    The Competent Authority has a statutory duty to consider the environmental information before granting consent to any project subject to the EIA process.

**D.9.2**    PAN 58 provides useful guidance on this stage. Paragraphs 80–90 discuss the process of evaluation of the Environmental Statement and its review. Annexe 5 provides a checklist of 'quality indicators' and five headings under which a Statement may be reviewed:

a.  elements of the project            d.  mitigating measures
b.  policy framework                   e.  risks and hazardous development.
c.  environmental effects

## Reviewing Environmental Statements

**D.9.3**    In addition to the advice in the Circular and PAN 58, this Handbook includes Attachment A which is a review package for the scoping and reviewing of the Environmental Statement stages in the EIA process. These are intended to be helpful working tools for adaptation by users to meet particular circumstances. They will hopefully assist in a more systematic and logical approach to these stages for EIA. They are not intended either to replace any existing formal review procedures undertaken by Competent Authorities or consultees, or to establish inflexible or standardised approaches to good practice. Users are positively encouraged to extend, reduce or otherwise adapt the frameworks suggested to suit particular needs.

# D.10   Formulating a Consultation Response

| | Step in the EIA Process |
|---|---|
| **Stage 1:**<br>**Before Submission of the**<br>**Environmental Statement** | Deciding whether EIA is required<br>Requiring submission of an Environmental Statement<br>Preliminary contacts and liaison<br>Scoping the Environmental Statement<br>Information collection<br>Describing baseline environmental information<br>Predicting environmental impacts<br>Assessing the significance of impacts<br>Mitigation measures and enhancement<br>Presenting environmental information in the Environmental Statement |
| **Stage 2:**<br>**Submission of Environmental**<br>**Statement and Consideration of**<br>**Environmental Information** | Submission of Environmental Statement and project application for consent<br>Consultation and publicity<br>Requiring more information<br>Negotiating modifications to the project<br>Considering the environmental information |
| **Stage 3:**<br>**Making the Decision** | Making the decision<br>Guaranteeing compliance |
| **Stage 4:**<br>**Implementation** | Implementation of mitigation and compensation measures<br>Monitoring<br>Review, reassessment and remedial measures<br>Reporting |

**[See also Sections C.9, D.9, E.1, E.2, E.3 and E.4, Appendices 1–6, Attachment A and Annexe 2]**

## Consultee's Role

**D.10.1**   Consultees will review the Environmental Statement and comment on the application for the proposal. Consultees may assist the Competent Authority and advise on the adequacy and conclusions of the environmental information.

## Statutory Provisions and Government Guidance

**D.10.2**   The Competent Authority has a statutory duty to consider the environmental information before granting consent to any project subject to the EIA process (Reg. 3 EIASR 99). Consultees should also provide advice  to the Competent Authority on matters within their remit, where advice is requested. A consultee's response is a part of the environmental information that the Competent Authority must consider (Reg. 2 EIASR 99).

## The Consultation Response

**D.10.3**   Whilst the consultee's comments on the Environmental Statement and the letter making representations about the project itself are separate things, the representations about the acceptability of the project will clearly be informed and supplemented by the information in and comments on the Statement. Reference is made to section C.9 above, relating to requests for mitigation, even if the project, in principle, is acceptable.

**Box D.10.1**

**The Environmental Information**

It should be stressed that the environmental information is not just the Environmental Statement submitted by the developer, but also any additional information submitted by the developer and the comments of the statutory consultees and the public when received by the Competent Authority.

**D.10.4**   Therefore, the comments of a consultee should cover matters which it considers important which have been omitted from the Environmental Statement, as well as those which have been covered by the document. All of this information must be considered by the Competent Authority, and should be material to their decision. Indeed, research (25) and (32) found that responses by consultees were usually given more weight in the Competent Authority's decision than the Environmental Statement on which they were based.

★ Key advice ★

**Box D.10.1**
**Representations**

Comments should cover the following points:

- the accuracy of the Environmental Statement (especially baseline information and the prediction of impacts);

- the coverage of the Environmental Statement—whether there are important omissions, and whether the emphasis on the different impacts is appropriate;

- with respect to omissions of matters which the consultee considers to be important: the issues involved and further work required to address them;

- the level of confidence that the consultee has in the findings (i.e. the degree of uncertainty);

- whether the consultee agrees with the evaluation of significance of the impacts identified;

- whether the mitigating measures are satisfactory or not; and

- the adequacy of proposals in the Environmental Statement for monitoring impacts and responding to them.

In cases where the Environmental Statement is of a particularly poor quality, it may be appropriate for the consultee to make only a general, not a detailed response.

**D.10.5**   The consultee should provide its own evaluation of the importance of impacts. This should address whether the affected resource is of international, national, regional or local importance, and the degree to which the impact will affect the resource.

**D.10.6** It should be noted that the *comments* on the contents of an Environmental Statement are, technically speaking, distinct from the consultee's formal *response* to the application for development consent (e.g. planning application). The consultee's comments on the *Environmental Statement* are considered to be environmental information which informs the authority in its decision, whereas the response to the *application* is the consultee's view as to the best course of action available to the authority and the extent to which this view is, or is not, supported by the Environmental Statement.

**D.10.7** Thus, the consultee's comments on an Environmental Statement might be to the effect that the Environmental Statement accurately describes the impacts of a development, that the consultee agrees with the Environmental Statement that these impacts are significant and that the mitigation measures proposed in the Environmental Statement would not adequately address these impacts, although a modification of them would do so. The consultee's response to the application would therefore be that it objects to the development because of the significant natural heritage impacts detailed in the Environmental Statement, but would be minded to lift this objection if the suggested modified mitigation measures were incorporated into the conditions for the consent.

★ **Key advice** ★

**Box D.10.2**
**Response to Consultations: The Project and the Environmental Statement**

It is advisable to distinguish clearly between the 2 parts of a consultee's response to an application, by stating the formal response to the application in a covering letter, and appending comments on the Environmental Statement in an annexe.

## D.11 Outline Planning Applications

### Introduction

**D.11.1**  This section is intended to draw together all of the commentary and advice about outline planning applications and EIA that is found elsewhere in this Handbook. It therefore contains no new or different material from that found in the sections B.4, C.4 and D.6 above.

### Applying the EIA Regulations to Outline Planning Applications

**D.11.2**  Where it applies, the Directive requires EIA to be carried out prior to the grant of 'development consent'. Development consent is defined as 'the decision of the Competent Authority or Authorities which entitled the developer to proceed with the development'. Under the UK planning system, it is the planning permission that enables the applicant to proceed with the development. Therefore, where EIA is required for a planning application made in outline, the requirements of the Regulations must be fully met at the outline stage since reserved matters cannot be subject to EIA.

**D.11.3**  The planning permission and the conditions attached to it must be designed to prevent the development from taking a form—and having effects—different from what was considered during EIA. This was confirmed in the case of **R v SSTLR ex parte Diane Barker** (2001).

**D.11.4**  The cases of **R v Rochdale MBC ex parte Tew** (1999) and **R v Rochdale MBC ex parte Milne** (2000) set out the approach that planning authorities need to take when considering EIA in the context of an application for outline planning permission if they are to comply with the Directive and the Regulations. Both cases dealt with a legal challenge to a decision of the authority to grant outline planning permission for a business park. In both cases an Environmental Statement was provided. In **ex parte Tew** the Court upheld a challenge to the decision and quashed the planning permission. In **ex parte Milne**, the Court rejected the challenge and upheld the authority's decision to grant planning permission.

**D.11.5**  In **ex parte Tew**, the authority authorised a scheme based on an illustrative masterplan showing how the development might be developed, but with all details left to reserved matters. The Environmental Statement assessed the likely environmental effects of the scheme by reference to the illustrative masterplan. However, there was no requirement for the scheme to be developed in accordance with the masterplan and in fact a very different scheme could have been built, the environmental effects of which would not have been properly assessed. The Court held that description of the scheme was not sufficient to enable the main effects of the scheme to be properly assessed, in breach of Schedule 4 of the Regulations.

**D.11.6**  In **ex parte Milne**, the Environmental Statement was more detailed; a Schedule of Development set out the details of the buildings and likely environmental effects, and the masterplan was no longer merely illustrative. Conditions were attached to the permission *'to tie the outline permission for the business park to the documents which comprise the application'*. The outline permission was restricted so that the development that could take place would have to be within the parameters of the matters assessed in the Environmental

Statement. Reserved matters would be restricted to matters that had previously been assessed in the Environmental Statement. Any application for approval of reserved matters that went beyond the parameters of the Environmental Statement would be unlawful, as the possible environmental effects would not have been assessed prior to approval.

**D.11.7**  The judge emphasised that the Directive and Regulations required the permission to be granted in the full knowledge of the likely significant effects on the environment. This did not mean that developers would have no flexibility in developing a scheme. But such flexibility would have to be properly assessed and taken into account prior to granting outline planning permission.

**D.11.8**  He also commented that the Environmental Statement need not contain information about every single environmental effect. The Directive refers only to those that are likely and significant. To ensure it complied with the Directive the authority would have to ensure that these were identified and assessed before it could grant planning permission.

**D.11.9**  The Court of Appeal in **ex parte Diane Barker** (2001) confirmed this approach and there are some general conclusions that can be drawn about applications for outline planning permission:

a.  An application for a 'bare' outline permission with all matters reserved for later approval is extremely unlikely to comply with the requirement of the Regulations.

b.  When granting outline consent, the permission must be 'tied' to the environmental information provided in the Environmental Statement, and considered and assessed by the authority prior to approval. This can usually be done by conditions although it would also be possible to achieve this by a planning agreement (under section 75 of the Town and Country Planning (Scotland) Act 1997).

c.  An example of a condition was referred to in **ex parte Milne** (2000). *'The development on this site shall be carried out in substantial accordance with the layout included within the Development Framework document submitted as part of the application and shown on (a) drawing entitled "Master Plan with Building Layouts."* The reason for this condition was given as *'The layout of the proposed Business Park is the subject of an Environmental Impact Assessment and any material alteration to the layout may have an impact which has not been assessed by that process'* (see paras 28 and 131 of the judgement).

d.  Developers are not precluded from having a degree of flexibility in how a scheme may be developed. But each option will need to have been properly assessed and be within the remit of the outline permission.

e.  Development carried out pursuant to a reserved matters consent granted for a matter that does not fall within the remit of the outline consent will be unlawful.

## Scoping an Outline Planning Application

**D.11.10**  Where outline planning permission is sought, it may not be possible to predict impacts on the natural heritage at this stage, because the details of the development are insufficiently described or unknown. Therefore, the information in the EIA can only inform the decision in principle on whether development of the nature proposed is acceptable at all on the site.

**D.11.11** Circular 15/1999, paragraph 48 provides the following advice on outline applications as follows:

*Where EIA is required for a planning application made in outline, the requirements of the Regulations must be fully met at the outline stage since reserved matters cannot be subject to EIA. When any planning application is made in outline, the planning authority will need to satisfy themselves that they have sufficient information available on the environmental effects of the proposal to enable them to determine whether or not planning permission should be granted in principle. In cases where the Regulations require more information on the environmental effects for the Environmental Statement than has been provided in an outline application, authorities should request further information under regulation 19. This may also constitute a request under article 4(3) of the GDPO.*

## Requesting Further Information for Outline Planning Applications

**D.11.12** When any planning application is made in outline, the planning authority will need to satisfy themselves that they have sufficient information available on the environmental effects of the proposal to enable them to determine whether or not planning permission should be granted in principle. In cases where more information is required, authorities should request further information on the Environmental Statement as described below.

**D.11.13** Where outline planning permission is sought, it may not be possible to predict impacts on the environment with the information submitted, because the details of the development are unknown. Therefore, the information in the EIA can only inform the decision in principle on whether development of the nature proposed is acceptable at all on the site. If the information available in the Environmental Statement at this stage is insufficient to determine whether the development is acceptable in principle, the planning authority should require such information to be submitted as it is reasonably necessary to assess the likely environmental effects of the proposal or they should refuse planning permission, possibly with an indication that a detailed application would be considered if it is supported by an Environmental Statement. (See also para 48 of Circular 15/1999.)

**D.11.14** In respect of all planning applications, the planning authority has the power to require information to be submitted under 2 statutory provisions, namely:

a) Regulation 19 of the EIASR 99, requiring submission of:

i) any further environmental information to enable the application to be determined; or

ii) information concerning any matter which is required to be dealt with in the Environmental Statement (i.e. matters in Schedule 4 EIASR 99); or

iii) information reasonably required to give proper consideration to the application; or

iv) evidence to verify any information in the Environmental Statement;

b) Article 13 of the Town and Country Planning (General Development Procedure) (Scotland) Order 1992 requiring any further information in order to enable them to deal with the application.

**D.11.15**   In respect of outline planning applications, the planning authority has the power to require reserved matters to be submitted under the provisions of Article 4(3) of the Town and Country Planning (General Development Procedure) (Scotland) Order 1992. This enables a planning authority to require the submission of all or any reserved matters that the planning authority considers it to be necessary to consider before the grant of an outline planning permission. However, whereas there is no limit to the period in which the other provisions can be used to require information to be submitted, there is a 1 month time limit on the use of Article 4(3) requiring reserved matters to be submitted.

**D.11.16**   **Thus, requests for some of the reserved matters to be submitted must be made by the planning authority within 1 month under Article 4(3) of the GDPO; requests for further information about the proposal that forms the subject of the planning application should be obtained (at any time) by means of Article 13 of the GDPO; and further information on the environmental impacts of the proposal should be obtained (at any time) by means of Regulation 13 of the EIASR 99.**

**D.11.17**   Which of the reserved matters a consultee needs to have addressed by the planning authority before it can reasonably determine the application is dependant on the nature of the proposal and the nature of the environmental sensitivities of the site. A proposal on or near a bog or mire, for instance, will require details of such reserved matters as access and road drainage so that their hydrological effects can be assessed. Where there are landscape and visual sensitivities, the siting, mass and height of the main components of the development, and possibly ancillary development such as roads, car parks, etc., will be necessary. Each proposal, however, is unique and will have to be considered carefully.

# Part E

# The Decision Making Stage

# E.1   Adopting the Precautionary Principle

| | Step in the EIA Process |
|---|---|
| **Stage 1:** <br> **Before Submission of the** <br> **Environmental Statement** | Deciding whether EIA is required <br> Requiring submission of an Environmental Statement <br> Preliminary contacts and liaison <br> Scoping the Environmental Statement <br> Information collection <br> Describing baseline environmental information <br> Predicting environmental impacts <br> Assessing the significance of impacts <br> Mitigation measures and enhancement <br> Presenting environmental information in the Environmental Statement |
| **Stage 2:** <br> **Submission of Environmental** <br> **Statement and Consideration of** <br> Environmental Information | Submission of Environmental Statement and project application for consent <br> Consultation and publicity <br> Requiring more information <br> Negotiating modifications to the project <br> Considering the environmental information |
| Stage 3: <br> **Making the Decision** | **Making the decision** <br> Guaranteeing compliance |
| **Stage 4:** <br> **Implementation** | Implementation of mitigation and compensation measures <br> Monitoring <br> Review, reassessment and remedial measures <br> Reporting |

**[See also Sections D.6–D.10, E.2, E.3 and E.4, Attachment A and Annexe 2]**

## The Precautionary Principle

**E.1.1**   This principle is particularly relevant to the EIA process. Generally, decisions should be based on the best scientific and other information available.

**E.1.2**   The EIA should ensure that this is available to the decision maker, at the right time. The environmental information should make clear, or as clear as possible, the environmental effects and consequences of the project. However, there are bound to be limitations in many cases where prediction is uncertain, e.g. based largely on professional judgement using assumptions that themselves are uncertain. Comparison with the effects of other projects elsewhere is often not available and sometimes it is not practical or feasible to obtain all the information desirable, e.g. where considerable costs or long time scales are involved.

**E.1.3**   The principle was described in the Rio Declaration 1992 which set out the 'precautionary approach':

*Where there are threats of serious or irreversible damage, lack of full scientific certainty shall not be used as a reason for postponing cost-effective measures to prevent environmental degradation.*

**E.1.4**   This wording indicated that the principle can be applied to all forms of environmental damage that might arise and should not be confined only to the actions of government.

**E.1.5**    Importantly, the precautionary principle is addressed in some detail in PAN 58. At paragraph 94:

**The precautionary principle** – *the principle that authorities should act prudently to avoid the possibility of irreversible environmental damage in situations where the scientific evidence is inconclusive but the potential damage could be significant.*

*It applies particularly where there are good grounds for judging either that action taken promptly at comparatively low cost may avoid more costly damage later, or that irreversible effects may follow if action is delayed.*

**E.1.6**    NPPG 14 Natural Heritage at paragraphs 80–82 states *'planning authorities should apply the precautionary principle in circumstances where the impacts of a proposed development are uncertain, but there are good grounds for believing that significant irreversible damage could occur to natural heritage interests of international or national significance. Where it appears that a precautionary approach is justified, careful consideration should be given to whether the proposal might be modified to eliminate the risk of irreversible damage before a decision is reached to refuse planning permission.'*

**E.1.7**    In cases where an internationally designated nature conservation site may be affected, Regulation 48 of the *Conservation (Natural Habitats &c) Regulations* 1994 embodies the precautionary principle in the requirement to grant consent (subject to the derogations in Regulation 49) only if the Competent Authority has ascertained that the project will not adversely affect the integrity of the site. There is no requirement to demonstrate that there would be harm, the duty is to establish that there would be no harm to the integrity of the site.

★ **Key advice** ★

**Box E.1.1**
**The Precautionary Approach**

Competent Authorities should adopt the precautionary approach in considering environmental information and when deciding whether to consent to projects, in accordance with Government policy.

**E.1.8**    The SNH approach and recommendations as to the application of the precautionary principle are set out in *Applying the Precautionary Principle to decisions on the natural heritage*, 2001, SNH.

## E.2    Relationship of EIA with the Development Plan and Other Consent Proceedures

| | Step in the EIA Process |
|---|---|
| **Stage 1:**<br>**Before Submission of the**<br>**Environmental Statement** | Deciding whether EIA is required<br>Requiring submission of an Environmental Statement<br>Preliminary contacts and liaison<br>Scoping the Environmental Statement<br>Information collection<br>Describing baseline environmental information<br>Predicting environmental impacts<br>Assessing the significance of impacts<br>Mitigation measures and enhancement<br>Presenting environmental information in the Environmental Statement |
| **Stage 2:**<br>**Submission of Environmental**<br>**Statement and Consideration of**<br>**Environmental Information** | Submission of Environmental Statement and project application for consent<br>Consultation and publicity<br>Requiring more information<br>Negotiating modifications to the project<br>Considering the environmental information |
| **Stage 3:**<br>**Making the Decision** | **Making the decision**<br>Guaranteeing compliance |
| **Stage 4:**<br>**Implementation** | Implementation of mitigation and compensation measures<br>Monitoring<br>Review, reassessment and remedial measures<br>Reporting |

**[See also Sections D.6–D.10, E.1, E.3 and E.4 and Attachment A]**

### Role of EIA

**E.2.1**    It is important to bear in mind that the EIA process is only one part of the decision making procedure and that the Environmental Statement is only one part of the EIA process. Guidance on this issue is also provided in PAN 58 at paragraphs  10–15.

### Planning Related Decisions

**E.2.2**    For example, when dealing with a planning application a planning authority must decide the application in accordance with the development plan unless material considerations indicate otherwise (see TCPSA 1997 S.25). The environmental information is a material consideration. The Environmental Statement is an important part of the environmental information. There is no requirement for the planning authority or Reporter or Scottish Ministers to agree with or to adopt or reject the conclusions of an Environmental Statement. They need to take it into account and, if granting permission, to state in their decision that they have taken the environmental information into account (Reg. 3 EIASR 99).

**E.2.3**    Environmental Statements relating to development requiring planning permission should directly relate the environmental effects of the project to the relevant development plan policies: all of them, not just a favourable selection. It should be clear from the Environmental Statement whether the development is in accordance with the development plan. Whether or not it is in accordance with the development plan, it is open to the developer in the Environmental Statement or in his submissions explaining the proposals to the planning authority, what other material considerations may be relevant to the planning decision.

**E.2.4**     Consequently, an Environmental Statement may fairly conclude that the project is not in accordance with some development plan policies because of its adverse environmental effects but, nevertheless, the Environmental Statement may set out material considerations which could outweigh the policies–such as economic benefits or benefits to other aspects of the environment that may be enhanced rather than harmed.

## General Principles

**E.2.5**     These same principles apply to all competent authorities and all decision making procedures. EIA is intended to inform the decision not to direct what decision should be made.

## Internationally Designated Nature Conservation Sites (Natura Sites)

**E.2.6**     If a project would be likely to have a significant effect on a Natura 2000 site in Great Britain, and it is not necessary for the management of that site, then the decision maker must follow the procedures in Regulations 48 and 49 of the Habitats Regulations 1994 and carry out an appropriate assessment.

**E.2.7**     The appropriate assessment is not the same as an EIA under the provisions of the EIA Regulations. Compliance with the Directives 85/337/EEC and 97/11/EC is achieved through the EIA process which should run alongside and concurrently with the 'appropriate assessment' under the Habitats Regulations in compliance with Directive 92/43/EEC. Neither procedure overrides the other; both must be followed where both sets of Regulations apply. In many cases, plans or projects that will be subject to an appropriate assessment will need an Environmental Statement to be prepared under the EIA Regulations.

**E.2.8**     The Environmental Statement will address all significant environmental effects. The appropriate assessment will only address the effects of the proposal on the internationally important habitats and/or species for which the site is or will be designated or classified. It will be appropriate to use the information assembled for the Environmental Statement when carrying out the appropriate assessment under the Habitats Regulations (Circular 6/1995 as modified by SEERAD in 2000, Annexe D Appendix A paragraph 3 (35)). In view of this it would be helpful if relevant Environmental Statements clearly identified, under a specific heading, the likely significant effects on the internationally important habitats and/or species.

**E.2.9**     It should also be noted that, in Natura 2000 site casework, the consideration as to whether the proposal would be likely to have a significant effect on the site is to be made in view of the site's conservation objectives. These should be provided to the developer by SNH at the earliest opportunity in relevant cases. The developer should seek guidance on the assessment from SNH. If the information for the appropriate assessment under Regulation 48 is to be included in the Environmental Statement, it should include an assessment of each of the site's (international) interest features in view of the conservation objectives for those interests.

**E.2.10**   The other main implication of the Habitats Regulations is that there is a greater need for the developer to consider and set out alternative solutions, showing why there are none or why they must be rejected, so that the Competent Authority may determine whether there are alternative solutions under the procedures in Regulation 49 of the Habitats Regulations, should it be necessary to apply the requirements of Regulation 49.

### Influence of the EIA Process

**E.2.11**   Research (25) has shown that,` with increasing experience of EIA, Environmental Statements have become more open and well balanced and therefore a more credible part of the decision making process. Environmental Statements completed since 1992, by experienced assessors, demonstrated a more objective, impartial and rigorous approach.

**E.2.12**   The EIA process can be extremely influential. Even where decision making authorities are inexperienced in the EIA process, or they have no expertise in some aspects of the assessment, they generally treat the process seriously and some seek expert advice and guidance where necessary. However, this is sometimes constrained by a lack of resources to commission external help.

---

★ **Key advice** ★

**Box E.2**
**Consultation Responses**

It is vital that consultees concentrate on making representations about the project—clearly setting out their opinion as to the effects on the environment and the significance of the effects, and where appropriate, whether the proposal should be given consent or other authorisation.

These representations can, and should, draw upon the information in the Environmental Statement and indicate whether the conclusions in the Environmental Statement are a sound basis for informing the Competent Authority as to the effects on the environment.

The response should not, however, focus entirely on the strengths or weaknesses of the Environmental Statement.

Detailed comments on the Environmental Statement may assist the Competent Authority and may be important, but the consultee's response should clearly distinguish between the formal response to the application, which should be in the covering letter, and the comments on the Environmental Statement, which might usefully be included in an Annexe to the consultee's main response.

---

**E.2.13**   Expert advice and guidance usually comes from statutory consultees or other well-informed commentators. Generally, the comments of these bodies are considered carefully and weight is attached to Environmental Statements, which the consultees consider to be well prepared, balanced and competent. It follows that statements prepared in the erroneous belief that they can be used to conceal adverse impacts and promote alleged environmental enhancement are not given weight in the decision. Poorly balanced or ill prepared statements can form an obstacle to the decision. They have led to scepticism, lack of credibility, delay and often a refusal of the consent being sought.

**E.2.14**   Well balanced, thoroughly prepared, clear and comprehensive statements expedite the decision making process, reduce the need to apply precautionary restrictions and increase confidence that the project would be responsibly undertaken with a commitment to mitigation.

**E.2.15**   The influence of the consultees, both statutory and non-statutory, is vital to the process. Clearly specified and reasoned requests for scoping, survey information, analysis, prediction and mitigation are usually received positively by decision makers and developers. As a result of consultation responses, Environmental Statements are frequently improved or supplemented, the effect of mitigating measures enhanced and projects modified.

## E.3 Guaranteeing Commitments and Compliance

| | Step in the EIA Process |
|---|---|
| **Stage 1:**<br>**Before Submission of the**<br>**Environmental Statement** | Deciding whether EIA is required<br>Requiring submission of an Environmental Statement<br>Preliminary contacts and liaison<br>Scoping the Environmental Statement<br>Information collection<br>Describing baseline environmental information<br>Predicting environmental impacts<br>Assessing the significance of impacts<br>Mitigation measures and enhancement<br>Presenting environmental information in the Environmental Statement |
| **Stage 2:**<br>**Submission of Environmental**<br>**Statement and Consideration of**<br>Environmental Information | Submission of Environmental Statement and project application for consent<br>Consultation and publicity<br>Requiring more information<br>Negotiating modifications to the project<br>Considering the environmental information |
| Stage 3:<br>Making the Decision | Making the decision<br>**Guaranteeing compliance** |
| Stage 4:<br>**Implementation** | Implementation of mitigation and compensation measures<br>Monitoring<br>Review, reassessment and remedial measures<br>Reporting |

### [See also Sections D.6–D.10, E.1, E.2, and E.4 and Attachment A]

### Statutory Provisions and Government Guidance

**E.3.1**    The Competent Authority has statutory powers to impose conditions, restrictions or limitations on the project consent and/or to enter into legal agreements to guarantee compliance with the terms of the consent. Circular 15/1999 strongly endorses the approach in this section of the handbook at paragraphs 123–127. Guidance on this stage is also provided in PAN 58 at paragraphs 55–61 and 93–97.

### Conditions and Other Limitations

**E.3.2**    The granting of consent for a project almost always relies on conditions that are intended to limit or restrict the development and on the implementation of the mitigating measures. Without the conditions and the mitigation the project would be environmentally unacceptable.

**E.3.3**    However, the Directive and the Regulations do not require the implementation of the mitigation measures specified in the Environmental Statement or elsewhere. The implementation and enforcement is left to the consenting procedures.

**E.3.4**    It is not sufficient, therefore, for an Environmental Statement merely to indicate what the mitigating measures would be. They must each be clearly identified (a statutory requirement of the Regulations, see Section B.6 above); and should be guaranteed in the event of the project proceeding. Neither is it likely to be sufficient for a condition on a consent which merely states that the development shall be *'in accordance with the environmental statement'*; Circular 15/1999 at para 124 says this is likely to be too vague.

**E.3.5**    The Environmental Statement and/or the decision notice should expressly state how the various measures will be implemented. These may include, for example, requirements of conditions on planning permissions and licences or legally binding agreements.

**E.3.6**    The usual form of obligation for projects subject to the town and country planning procedures are planning agreements under Section 75 of the Town and Country Planning (Scotland) Act 1997 (formerly S.50 Agreements under the 1972 Act). These agreements, which may be binding on successors in title, are enforceable by the planning authority and have a good record of compliance which provides confidence for the public and interested bodies.

**E.3.7**    Alternatively, Circular 15/1999 at paragraph 127 urges consideration of developers adopting environmental management systems such as the Eco Management and Audit Scheme (EMAS) to demonstrate implementation of mitigation measures and to monitor their effectiveness. However, the wording of this paragraph clearly indicates that the Scottish Executive sees this as 'In addition' to the conditions and agreements described above.

★ **Key advice** ★

**Box E.3**
**Conditions and Agreements**

In order for mitigation measures proposed in the Environmental Statement to be binding, they must form part of the application, conditions of consent, or other legal agreement (e.g. Section 75 Planning Agreement) between the Competent Authority and the developer.

Monitoring impacts should be covered by a Section 75 Agreement, or equivalent. Therefore, Competent Authorities and consultees should ensure that appropriate provisions are made in the consent.

## E.4  The Decision of the Competent Authority

| Step in the EIA Process | |
|---|---|
| **Stage 1:**<br>**Before Submission of the**<br>**Environmental Statement** | Deciding whether EIA is required<br>Requiring submission of an Environmental Statement<br>Preliminary contacts and liaison<br>Scoping the Environmental Statement<br>Information collection<br>Describing baseline environmental information<br>Predicting environmental impacts<br>Assessing the significance of impacts<br>Mitigation measures and enhancement<br>Presenting environmental information in the Environmental Statement |
| **Stage 2:**<br>**Submission of Environmental**<br>**Statement and Consideration of**<br>**Environmental Information** | Submission of Environmental Statement and project application for consent<br>Consultation and publicity<br>Requiring more information<br>Negotiating modifications to the project<br>Considering the environmental information |
| **Stage 3:**<br>**Making the Decision** | **Making the decision**<br>Guaranteeing compliance |
| **Stage 4:**<br>**Implementation** | Implementation of mitigation and compensation measures<br>Monitoring<br>Review, reassessment and remedial measures<br>Reporting |

**[See also Figure 2, Sections D.6–D.10, E.1, E.2 and E.3 and Attachment A]**

### Statutory Provisions and Government Guidance

**E.4.1**    The Competent Authority must state in writing, when granting a consent to a project that was subject to EIA, that the environmental information has been taken into account. (Regulations 3 EIASR 99 and para 122 Circular 15/1999. Guidance on this stage is also provided in PAN 58 at paragraphs 55–57 and 93–97.)

**E.4.2**    The Competent Authority must also notify the Scottish Ministers, the consultation bodies and the applicant of their decision, irrespective of whether they are granting or refusing the consent. They must also publicise their decision in the local press and indicate in the press notice where a copy of the decision making documents and the decision are available for public inspection, free of charge (Reg. 21 EIASR 99 and paragraphs 128–130 Circular 15/1999).

**E.4.3**    Consultees should advise the Competent Authority on matters affecting their remit, where advice is requested. **The Competent Authority is required to notify the consultation bodies of the decision on the project application** (Reg. 21 EIASR 99). This is often overlooked. Para 128 of Circular 15/1999 only refers to the requirement to notify the Scottish Ministers and the applicant and the public press notice; there is no reference to the consultation bodies. However, the Regulations are clear at Reg. 21(1)(a) that statutory consultees must be notified.

## The Decision

**E.4.4**     The Competent Authority will make its decision on whether to consent to the project. The Regulations require that the environmental information must be taken into account. There is no duty on the Competent Authority to agree with the conclusions of the Environmental Statement or to accept the advice or recommendations of the consultees or the public. The duty is limited to taking all of the information into account. It is, therefore, open to the Competent Authority to grant consent to an environmentally damaging project or to refuse consent for an environmentally beneficial or benign project.

**E.4.5**     The Competent Authority must state on the face of the consent that they have taken account of the environmental information, in accordance with the Regulations. They do not have to do this if they are refusing consent. Indeed, if refusing consent they do not have to take the environmental information into account, in order to comply with the Regulations, although they almost certainly will do to give further and better reasons for refusing consent. They are bound to notify the Scottish Ministers and the consultation bodies of their decision, whether or not they grant permission.

**E.4.6**     For planning applications, a copy of the decision, including any conditions imposed, must be kept with the planning register and along with such other documents as contain:

a.  the main reasons and considerations on which the decision was based; and

b.  where permission has been granted, a description of the main measures to avoid, reduce and, if possible, offset the major adverse effects of the development.

Circular 15/1999 fairly indicates that in most cases a copy of the planning officer's report to the committee is likely to meet these requirements.

## Projects Using Uncultivated Land and Semi-Natural Areas for Intensive Agriculture (ULSNA)

**E.4.7**     Under the provisions of the ULSNAR02 SEERAD can refuse to grant consent, in which case the project would not be able to proceed, subject to the appeal procedures described in section B.5 above. SEERAD may grant consent, with or without conditions. Any conditions on a consent must be complied with or an offence will be committed.

# Part F

# Implementation and Compliance

## F.1 Implementation of Mitigation and Compensation Measures

**For each of the pre-construction, construction, operational, decommissioning and restoration stages**

| | Step in the EIA Process |
|---|---|
| **Stage 1:** **Before Submission of the Environmental Statement** | Deciding whether EIA is required |
| | Requiring submission of an Environmental Statement |
| | Preliminary contacts and liaison |
| | Scoping the Environmental Statement |
| | Information collection |
| | Describing baseline environmental information |
| | Predicting environmental impacts |
| | Assessing the significance of impacts |
| | Mitigation measures and enhancement |
| | Presenting environmental information in the Environmental Statement |
| **Stage 2:** **Submission of Environmental Statement and Consideration of Environmental Information** | Submission of Environmental Statement and project application for consent |
| | Consultation and publicity |
| | Requiring more information |
| | Negotiating modifications to the project |
| | Considering the environmental information |
| **Stage 3:** **Making the Decision** | Making the decision |
| | Guaranteeing compliance |
| **Stage 4:** **Implementation** | **Implementation of mitigation and compensation measures** |
| | Monitoring |
| | Review, reassessment and remedial measures |
| | Reporting |

## [See also Sections C.9, D.6, D.9, D.10, E.2, E.3, E.4, F.2 and F.3]

### Statutory Provisions and Government Guidance

**F.1.1**    The developer has a statutory duty to comply with the terms of the consent. The Competent Authority has statutory powers to enforce compliance. Guidance on this stage is also provided in PAN 58 at paragraphs 55–61

**F.1.2**    There is no duty on the Competent Authority to monitor compliance with conditions and the terms of the consent. Enforcement will often rely on interested parties such as statutory consultees or local residents drawing any non-compliances to the attention of the Competent Authority. Consultees may not be made aware of the commencement of the project. The extent of monitoring for compliance with terms and conditions, which consultees relied on in the decision to grant consent, needs to be judged on a case-by-case basis, depending on the issues involved, the resources required and available, and the confidence in the Competent Authority and the developer.

### Implementation

**F.1.3**    Many Environmental Statements will contain a project programme indicating the likely start and end dates of the main phases of the project, assuming consent is granted. However, these are often over-optimistic as to the length of time it will take to obtain the consent. Such programmes may well be out of date by the time the consent is issued. Developers will usually be willing to advise consultees and the Competent Authority of any revisions to programmes, on request.

**F.1.4**    The degree of monitoring will vary according to the type of development and some phases may be more environmentally sensitive than others. Usually, the key phases will be site preparation and construction and, at a later date, decommissioning and/or restoration. Many schemes will include advance mitigation works, e.g. advanced planting for screening, and these may need to be checked.

★ **Key advice** ★

**Box F.1.1**
**The Approach to Monitoring**

It will need to be decided, on a case-by-case basis, which projects should be monitored for compliance, how such monitoring should be undertaken and by whom, and which of the mitigation measures should be checked, at which stages of the development.

Consultees should work closely with the competent authorities to draw up appropriate conditions and agreements to ensure adequate monitoring (quarterly, annually, etc., as appropriate to the nature of the concern) and provision for mitigation (which could include financial and other guarantees).

See also Sections F.2–3 below.

# F.2    Monitoring Programmes

| Step in the EIA Process | |
|---|---|
| **Stage 1:** **Before Submission of the Environmental Statement** | Deciding whether EIA is required |
| | Requiring submission of an Environmental Statement |
| | Preliminary contacts and liaison |
| | Scoping the Environmental Statement |
| | Information collection |
| | Describing baseline environmental information |
| | Predicting environmental iImpacts |
| | Assessing the significance of impacts |
| | Mitigation measures and enhancement |
| | Presenting environmental information in the Environmental Statement |
| **Stage 2:** **Submission of Environmental Statement and Consideration of Environmental Information** | Submission of Environmental Statement and project application for consent |
| | Consultation and publicity |
| | Requiring more information |
| | Negotiating modifications to the project |
| | Considering the environmental information |
| **Stage 3:** **Making the Decision** | Making the decision |
| | Guaranteeing compliance |
| **Stage 4:** **Implementation** | Implementation of mitigation and compensation measures |
| | **Monitoring** |
| | Review, reassessment and remedial measures |
| | Reporting |

## [See also Sections C.9, D.6, D.9, D.10, E.2, E.3, E.4, F.1 and F.3]

### Statutory Provisions and Government Guidance

**F.2.1**    Monitoring is a non-statutory procedure but may be required by conditions on a project consent, or by legal agreements (such as planning agreements under S.75 of the Town and Country Planning (Scotland) Act 1997), that would be legally enforceable by the Competent Authority. Guidance on this stage is also provided in PAN 58 at paragraphs 55–61.

### Monitoring

**F.2.2**    Implementation of mitigating measures may still not guarantee their success in reducing environmental effects. It is vital that someone monitors the effectiveness of mitigation to ensure that it meets the standards and achieves the objectives anticipated in the decision. Monitoring can improve the future mitigation of similar developments. It may also be necessary where no mitigation was proposed or required because the development was not expected to cause significant environmental change. The Directive and Regulations do not require monitoring procedures to be put in place, only mitigation measures.

**F.2.3**    Post-project monitoring and review are appropriate to planning and other legal agreements and should be clearly described and guaranteed in the EIA process. The Environmental Statement should contain a prescription for the implementation of mitigating measures, monitoring and review procedures with a clear commitment and readiness to accept conditions and legal agreements to ensure they are implemented at the right time and in appropriate ways.

**F.2.4** Consultees may be able to make a valuable contribution to the design of monitoring, and will have the opportunity, as a statutory consultee, to comment on the adequacy of monitoring proposals set out in the Environmental Statement.

**F.2.5** Monitoring may be delegated to a range of bodies, which commonly include the developers or their consultants or university research teams. However, monitoring will not usually be feasible unless it is financed by the developer.

**F.2.6** However, monitoring to verify the predictions of EIA has seldom been undertaken in Great Britain, though it may be possible to obtain data relevant to the topic where developments are situated in, or close to, sites where surveys are proceeding for other reasons.

**F.2.7** The lack of such monitoring is common to EIA in all parts of the world, and has been identified as one of the primary reasons for the low scientific reliability of many EIAs worldwide.

---

★ **Key advice** ★

**Box F.2.1**
**Monitoring Programmes/Agreements**

Consultees should enter into an agreement to assist and advise in drawing up the schedule and methodology for monitoring and should agree to assess the results of monitoring and to advise the Competent Authority and developer of these results.

Consultees should be consulted by the Competent Authority when it is considering whether to approve or amend mitigation schemes, wherever the effects on the natural heritage are potentially significant. It is for the Competent Authority to ensure (enforce) that these conditions, monitoring and mitigation, are met. If there is a timetable for receipt of details of monitoring and this is not met Consultees should alert the Competent Authority and press them to take action. Similarly, if there is a timetable for agreeing and implementing mitigation measures and this is not met, or consultees believe it is not being met, consultees should alert the authority or press them to take action.

## F.3 Review and Reassessment and Remedial Programmes

| | Step in the EIA Process |
|---|---|
| **Stage 1:**<br>**Before Submission of the**<br>**Environmental Statement** | Deciding whether EIA is required<br>Requiring submission of an Environmental Statement<br>Preliminary contacts and lLiaison<br>Scoping the Environmental Statement<br>Information collection<br>Describing baseline environmental information<br>Predicting environmental impacts<br>Assessing the significance of impacts<br>Mitigation measures and enhancement<br>Presenting environmental information in the Environmental Statement |
| **Stage 2:**<br>**Submission of Environmental**<br>**Statement and Consideration of**<br>**Environmental Information** | Submission of Environmental Statement and project application for consent<br>Consultation and publicity<br>Requiring more information<br>Negotiating modifications to the project<br>Considering the environmental information |
| **Stage 3:**<br>**Making the Decision** | Making the decision<br>Guaranteeing compliance |
| **Stage 4:**<br>**Implementation** | Implementation of mitigation and compensation measures<br>**Monitoring**<br>Review, reassessment and remedial measures<br>Reporting |

### [See also Sections C.9, D.6, D.9, D.10, E.2, E.3, E.4, F.1 and F.2]

### Statutory Provisions and Government Guidance

**F.3.1**   Review, reassessment and remedial measures are non-statutory procedures but may be required by conditions on a project consent, or by legal agreements (such as planning agreements under S.75 of the Town and Country Planning (Scotland) Act 1997), that would be legally enforceable by the Competent Authority. Guidance on this stage is also provided in PAN 58 at paragraphs 58–61.

### Review

**F.3.2**   Provision must be made at the decision making stage to ensure that changes or remedial (i.e. corrective) action can be implemented effectively and quickly if monitoring reveals problems. Procedures for monitoring and the review of mitigation after the project has commenced, and for as long as may be necessary, are therefore essential if monitoring is to have any real effect.

**F.3.3**   The key point about monitoring is that it should not be monitoring for its own sake.  There may be occasions when monitoring simply to verify or validate the predictions in the Environmental Statement may be appropriate (to assist predictions in other, similar cases in the future) but usually monitoring will only be worthwhile if it is reinforced with effective review and remedial action mechanisms. These may include reassessment of the project in the light of actual effects that occur, or may include observation and reporting on the nature and scale of effects and comparison with those predicted in the Environmental Statement.

**F.3.4**    Reviews may need to include consultation. Often this can be accommodated by an annual report (or some other appropriate time scale) being submitted to the Competent Authority and statutory consultees by the developer's consultants or the monitoring team. These reports could be considered at an annual review meeting where the relevant parties decide the effectiveness of the mitigation.

**F.3.5**    Again, review is only worthwhile where there is a clear purpose to it. If there are no mechanisms whereby the developer has agreed to adjust or otherwise change mitigation, in the light of the monitoring and review, then there is usually no point reviewing the monitoring.

---

★ **Key information** ★

**Box F.3.1**
**Guaranteeing Monitoring**

The decision of the Competent Authority in deciding to grant consent or authorisation for the project, or legally binding agreements drawn up at the time of the decision, should make clear what procedures will be put in place to review the monitoring and to change the mitigation if necessary.

They should indicate who will review the effects, who will report to whom, who is responsible for taking decisions, who will implement the changes to mitigation and other remedial works, and who will pay the costs of remedial work and corrective action. It is unlikely that these matters will be appropriate for inclusion in a planning condition, and a S.75 Agreement or similar legally enforceable agreement will normally be required.

# Annexes

**Alternative solutions**

are alternative ways of achieving the objectives of the project. They may include:

• alternative locations that are suitable and available; or

• different approaches in terms of design, manufacturing or other processes; the use of different forms of transport or energy; different sources for the supply of materials etc.

**Annexe I projects** (also referred to as Schedule 1 projects)

See Schedule 1 Projects below.

**Annexe II projects** (also referred to as Schedule 2 projects)

See Schedule 2 projects below.

**Competent Authority**

is the authority which determines the application for a consent, permission, licence or other authorisation to proceed with a development. It is the authority that must consider the environmental information before granting any kind of authorisation. For example, for projects requiring planning permission this will usually be the Planning Authority, but in some cases may be the Scottish Ministers, for Woodland Grant Scheme applications it is the Forestry Authority, for marine fish farms it is the Crown Estate Commissioners etc.

**Consultation bodies**

are any body specified in the relevant EIA Regulations which the Competent Authority must consult in respect of an Environmental Statement, and which also have a duty to provide information or advice during the EIA process. They are:

a. any adjoining planning authority, where the development is likely to affect land in their area;

b. Scottish Natural Heritage;

c. the water and sewerage authority or authorities for the area in which the development is to take place;

d. the Scottish Environment Protection Agency;

e. the Health and Safety Executive;

f. the Scottish Ministers.

**Crown Land/The Crown**

is a generic term for land held by Her Majesty the Queen as Monarch and certain other royal land and all Government held land, for example land held by the Ministry of Defence and land owned by the Scottish Ministers including prisons, Trunk Roads and Motorways.

**Developer**

For the purposes of this Handbook, to help make the text more readable, all project proposers are referred to as 'developers', whether or not their project constitutes development within the meaning of the Town and Country Planning (Scotland) Act 1997 and whether or not the project is for public service or infrastructure or for commercial purposes.

## Do-nothing comparison,

or in some cases, such as road improvements, the 'do-minimum' comparison, is a projection of the existing data to provide a baseline for comparison to show how the site would change if the project did not go ahead.

## EEA State

A State which is a Contracting Party to the Agreement on the European Economic Area signed at Oporto on 2nd May 1992 as adjusted by the Protocol signed at Brussels on 17th March 1993.

## EIA application

An application for planning permission for EIA development.

## EIA development

Development which is either:

a.  Schedule 1 development; or

b.  Schedule 2 development likely to have significant effects on the environment by virtue of factors such as its nature, size or location.

## Enhancement/Net Benefit/New Benefit

In natural heritage terms, this is the genuine enhancement of the natural heritage interest of a site or area because adverse effects are limited in scope and scale, and the project includes improved management or new habitats or features, which are better than the prospective management, or the habitats or features present there now. There is, therefore, a net or new benefit to the natural heritage.

## Environmental Impact Assessment

is the whole process of gathering environmental information; describing a development or other project; predicting and describing the environmental effects of the project; defining ways of avoiding, reducing or compensating for these effects; consulting the general public and specific bodies with responsibilities for the environment; taking all of this information into account before deciding whether to allow the project to proceed and ensuring that the measures prescribed to avoid, reduce or compensate for environmental effects are implemented.

## Environmental information

is the information that must be taken into account by the decision maker (the Competent Authority) before granting any kind of authorisation in any case where the EIA process applies. It includes the environmental statement, including any further information, any representations made by any body required by the Regulations to be invited to make representations, and any representations duly made by any other person about the environmental effects of the development;

## Environmental Statement

is the report normally produced by, or on behalf of, and at the expense of, the developer or project promoter which must be submitted with the application for whatever form of consent or other authorisation is required. It is only one component, albeit a very important one, of the environmental information that must be taken into account by the decision maker.

The EIASR 99 define it as a statement:

a.  that includes such of the information referred to in Part I of Schedule 4 as is reasonably required to assess the environmental effects of the development and which the applicant can, having regard in particular to current knowledge methods of assessment, reasonably be required to compile; but

b.  that includes at least the information referred to in Part II of Schedule 4.

## Exempt development

means development which comprises or forms part of a project serving national defence purposes or in respect of which the Scottish Ministers have made a direction under regulation 4(4).

## Iterative

(A process) repeated until the best solution has been found so, in the context of EIA, it can be understood as the process of assessment and reassessment until the best environmental fit is achieved.

## Mitigating measures or mitigation

are the measures taken to avoid, reduce or remedy adverse impacts of the project. They are:

*Avoidance*

which is the measures taken to avoid any adverse impacts, including alternative or 'do-nothing' options;

*Reduction*

which is the measures taken to reduce unavoidable adverse impacts of the project;

*Remedy or Compensatory measures or Compensation*

which are other measures taken to (at least try to) offset or compensate for residual adverse effects which cannot be avoided or further reduced.

## Revised Environmental Statement

Where a project has been modified since the original application and Environmental Statement were submitted, a revised Environmental Statement may be submitted, to amend the original, to ensure that the environmental information considered by the Competent Authority relates to the project as modified. The revised Environmental Statement may be a revision of the whole of the original document or revisions only of those parts of the original Environmental Statement that need to be changed as a result of the modifications.

## Schedule 1 projects

are plans or projects which are listed in Annexe I of the Directive, as revised, and Schedule 1 of the Regulations, as revised.

## Schedule 2 projects

are plans or projects which are listed in Schedule 2 of the Directive, as revised, and Schedule 2 of the Regulations, as revised.

## Schedule 1 application and Schedule 2 application

mean an application for planning permission for Schedule 1 development and Schedule 2 development respectively.

## Schedule 1 development

means development, other than exempt development, of a description mentioned in Schedule 1 of the EIASR 99.

## Schedule 2 development

means development, other than exempt development, of a description mentioned in Column 1 of the table in Schedule 2 of the EIASR 99 where:

a. any part of that development is to be carried out in a sensitive area; or

b. any applicable threshold or criterion in the corresponding part of Column 2 of that table is respectively exceeded or met in relation to that development.

## Scoping

is the procedure whereby the Competent Authority and the relevant statutory and other consultees are consulted at the outset, or very early in the EIA process, by the developer to agree what effects should be covered in the Environmental Statement, how they should be covered and the methods to be used to assess them. If requested by the developer the Competent Authority must give a scoping opinion.

## Screening

is the process of deciding whether a particular project that is proposed is EIA development, and therefore subject to the EIA process. It involves checking whether the project falls within the classes of project in Schedule 1 or 2 of the Regulations (or Annexe I or II of the Directives) and, if in Schedule 2, whether it would be likely to have significant effects on the environment.

## Screening direction

means a direction made by the Secretary of State as to whether development is EIA development.

## Screening opinion

means a written statement of the opinion of the relevant planning authority whether development is EIA development.

## Sensitive area

means any of the following:

a. a Site of Special Scientific Interest;

b. land to which S.23 of the Nature Conservation (Scotland) Act 2004 applies (Nature Conservation Areas);

c. a World Heritage Site (UNESCO 1972);

d. a schedule monument (Ancient Monuments and Archaeological Areas Act 1979);

e. a European site within the meaning of Reg. 10 of the Conservation (Natural Habitats, &c.) Regulations 1994 (SPA or SAC);

f. a National Scenic Area.

## Statutory consultee

is any body specified in the relevant EIA Regulations which the Competent Authority must consult in respect of an Environmental Statement, and which also has a duty to provide information or advice during the EIA process. They are listed in Section D.2 of this Handbook.

## Strategic Environmental Appraisal/Assessment (SEA)

the whole process of considering the environmental effects of plans, policies and proposed programmes of projects at a strategic level.

## Supplementary Environmental Statement

Where the original Environmental Statement was incomplete or further work on environmental effects has been undertaken (whether or not the project has been modified since the original application and Environmental Statement were submitted), a supplementary Environmental Statement may be submitted, to add to the original, to ensure that all of the relevant environmental information is considered by the Competent Authority. The supplementary Environmental Statement may include a revision of the whole or part of the original document or additions that are needed to cover the additional information.

| Legislation | Commentary/Description |
| --- | --- |
| **Parliamentary Standing Order No. 27A 20 May 1991 and General Order 27A 20 May 1992 (Inserted by the Private Legislation Procedure (Scotland) General Order 1992 (SI 1992 No. 1206))** | This Parliamentary Standing Order ensures that all Schedule 1 and Schedule 2 projects likely to have significant effects on the environment which are to be authorised by Parliament directly are subject to an EIA procedures, usually at Committee stage. However, this procedure is not a full EIA process and to date, the Scottish Parliament's Private Bill Committees for the National Galleries in Edinburgh, removing navigation rights to facilitate a wind farm in the Solway Firth, the Stirling–Alloa–Kincardine Railway, the Edinburgh Tram Systems and the Waverley Railway Line have not required developers to undertake the whole EIA process. |
| **Transport and Works Act 1992, S.14** | Ensures all Schedule 1 and Schedule 2 projects likely to have significant effects on the environment which are to be authorised or consented under the Transport and Works Act are subject to EIA. These projects may include a wide range of infrastructure works including roads, bridges, railways, light railway systems, harbour and port developments, inland navigation etc. |
| **Town and Country Planning (General Development Procedure) (Scotland) Order 1992 (SI 1992 No. 224)** | This General Development Order contains provisions for requiring further information on planning applications under Articles 6 and 13 (see Section D.6 of this Handbook); and for the Scottish Ministers to issue Directions about EIA under Articles 16 and 19. |
| **Transport and Works (Applications and Objections Procedure) Rules 1992 (SI 1992 No. 2902), Transport and Works (Assessment of Environmental Effects) Regulations 1995 (SI 1995 No. 1541), and Transport and Works (Assessment of Environmental Effects) Regulations 1998 (SI 1998 No. 2226)** | These regulations relate to the assessment of environmental effects of projects promoted via the Transport and Works Act 1992, for example for railways, tramways, inland waterways, bridges and works interfering with navigation. |
| **Town and Country Planning (Scotland) Act 1997, S.40** | Provides the Scottish Ministers with the power to make Regulations governing the EIA process generally, to add further types of projects to Schedule 2 of the Regulations and to make directions to planning authorities including whether an environmental statement should be submitted in any particular case. |

| Legislation | Commentary/Description |
|---|---|
| **Environmental Impact Assessment (Scotland) Regulations 1999 (Scottish Statutory Instrument 1999 No. 1)** | These Regulations cover EIA requirements for: <br>• decisions on planning applications, appeals and deemed planning permissions made under the Town and Country Planning (Scotland) Act 1997 (6) (Part II of the Regulations); <br>• certain trunk road projects, comprising construction and improvement which are authorised under the Roads (Scotland) Act 1984 (12) (Part III of the Regulations); <br>• agricultural drainage works authorised by the Scottish Ministers by way of an improvement order under the Land Drainage (Scotland) Act 1958 (Part IV of the Regulations). |
| **Environmental Impact Assessment (Forestry) (Scotland) Regulations 1999 (Scottish Statutory Instrument 1999 No. 43)** | EIA requirements for forestry works, including afforestation and reafforestation as regulated by the Forestry Commission through Grant Schemes and other measures under the Forestry Acts. |
| **Offshore Petroleum Production and Pipelines (Assessment of Environmental Effects) Regulations 1999 (SI 1999 No. 360)** | EIA regulations covering offshore oil industry and pipelines. |
| **Environmental Impact Assessment (Fish Farming in Marine Waters) Regulations 1999 (SI 1999 No. 367)** | These Regulations cover EIA requirements for fish farms in marine waters (fresh water fish farms would be subject to planning control and the Regulations in SSI 1999 No. 1, above). |
| **Public Gas Transporter Pipe-line Works (Environmental Impact Assessment) Regulations 1999 (SI 1999 No. 1672)** | EIA regulations for new gas pipelines and related infrastructure. |
| **Nuclear Reactors (Environmental Impact Assessment for Decommissioning) Regulations 1999 (SI 1999 No. 2892)** | EIA regulations for the decommissioning of nuclear reactors. |
| **Harbour Works (Environmental Impact Assessment) Regulations 1999 (SI 1999 No. 3445)** | EIA requirements for works undertaken by a Port or Harbour Authority under the provisions of the Merchant Shipping Act 1988 and the Harbours Act 1964 as regulated by the Scottish Ministers through Harbour Empowerment and Harbour Improvement Orders. |
| **Electricity Works (Environmental Impact Assessment) (Scotland) Regulations 2000 (Scottish Statutory Instrument 2000 No. 320)** | The Regulations relating to electricity power stations and overhead lines in Scotland. |
| **Pipeline Works (Environmental Impact Assessment) Regulations 2000 (SI 2000 No. 1928)** | These Regulations cover EIA requirements for pipeline projects in Scotland. |

| Legislation | Commentary/Description |
|---|---|
| **The Environmental Impact Assessment (Uncultivated Land and Semi-Natural Areas) (Scotland) Regulations 2002 (Scottish Statutory Instrument 2002 No. 6)** | A new regulatory mechanism for controlling projects that would use uncultivated land and semi-natural areas for intensive agriculture in order to ensure they were subject to EIA where necessary. |
| **The Environmental Impact Assessment (Scotland) Amendment Regulations 2002 (Scottish Statutory Instrument 2002 No. 324)** | Introduced the requirement to apply EIA procedures to the review of old mineral permissions in order to comply with a court ruling that the review and issue of new conditions amounts to the grant of a new consent that should be subject to EIA. |
| **The Electricity Act 1989 (Requirement of Consent for Offshore Generating Stations) (Scotland) Order 2002 (Scottish Statutory Instrument 2002 No. 407)** | Require all offshore generating stations mainly operated by wind or wave energy and over 1MW output subject to consenting procedures and the application of the Electricity Works (EIA) (Scotland) Regulations 2000. |
| **The Environmental Impact Assessment (Water Management) (Scotland) Regulations 2003 (Scottish Statutory Instrument 2003 No. 341)** | Amend the definition of development to include carrying out of irrigation or drainage or other water management works for agriculture so making such projects potentially EIA development subject to the EIASR 99. |
| **Town and Country Planning (Electronic Communications) (Scotland) Order 2004 (Scottish Statutory Instrument 2004 No. 332)** | Provides for the distribution of certain notices and other documents in administrative processes by email subject to caveats. |
| **The Environmental Information (Scotland) Regulations 2004 (Scottish Statutory Instrument 2004 No. 520)** | Transpose requirements of the Freedom of Information (Scotland) Act 2002 and EC Directive 90/313/EEC on public access to environmental information, requiring all public authorities to collect, maintain, disseminate and make available environmental information relevant to their functions. |
| **Environmental Impact Assessment and Habitats (Extraction of Minerals by Marine Dredging) Regulations (Draft)** | Draft. |

**The Application of EIA Regulations by Sector and Project Type**

| Sector | Project type | EIA Regulations |
|---|---|---|
| **Agriculture** | Buildings for intensive animal rearing<br>Development for irrigation and water management schemes<br>Land claim from the sea | EIASR 99 Part II |
| | Land drainage and flood prevention/control | EIASR 99 Part IV |
| | Conversion of uncultivated land to intensive agriculture | ULSNAR02 |
| | Abstraction of water, irrigation, drainage and other water management projects for agriculture | EIAWaterMR03 |
| **Aquaculture** | Installations for intensive marine fish farming | EIAFishFmMWR 99 |
| **Coastal projects** | Installations for intensive fresh water fish farming | EIASR 99 Part II |
| | Claiming land from the sea<br>Coast protection works<br>Flood banks and other flood prevention and control | EIASR 99 Part II |
| | Land drainage schemes | EIASR 99 Part IV |
| **Energy production and storage** | Coal, gas or oil fired power stations<br>Marine barrages for electricity generation<br>Nuclear power stations<br>Tidal and wave energy utilisation for electricity generation | ElecWorks EIASR 00 |
| | Development for hydro electric schemes<br>Development for wind turbine generators (wind farms)<br>Development for industrial briquetting of coal or lignite<br>Development for steam or hot water generation<br>Exploratory drilling for energy production<br>Geothermal drilling and utilisation<br>Oil refineries<br>Surface storage of natural gas and other fossil fuels<br>Underground storage of combustible gases | EIASR 99 Part II |
| **Energy transmission** | Offshore oil and gas production | OffshorePPPAEER 99 |
| | Offshore electricity generating stations | OffshoreGenStnsR02 |
| | Offshore oil and gas pipelines | OffshorePPPAEER 99 |
| | Overhead electricity power lines | ElecWorks EIASR 00 |
| | Gas pipelines | PGasTransPWEIAR 99 |
| | Oil/petroleum pipelines | PipelineWEIAR 00 |
| | Pipelines for transmission of steam or hot water | PipelineWEIAR 00 |

| Sector | Project type | EIA Regulations |
|---|---|---|
| **Energy decommiss-ioning and waste** | Decommissioning of nuclear power stations and other reactors | NuclearREIADR 99 |
| | Disposal of pulverised or other fuel ash<br>Drilling to store nuclear waste<br>Development for processing, reprocessing and storage of radioactive waste | EIASR 99 Part II |
| **Forestry** | Afforestation including natural regeneration<br>Deforestation<br>Forestry tracks and quarries | EIAForestrySR 99 |
| | Development for pulp/paper/board mills | EIASR 99 Part II |
| **Industrial development** | Development for all forms of industrial processing, reprocessing, manufacturing, assembling, packing, testing etc., and industrial estates | EIASR 99 Part II |
| **Leisure, sport and recreation** | Camping and caravanning sites<br>Golf courses and associated developments<br>Hotels, spas and similar complexes<br>Leisure centres<br>Marinas<br>Motor racing circuits and test tracks<br>Multiplex cinemas<br>Ski-runs, ski-lifts, cable cars, funicular railways<br>Sport stadiums<br>Theme parks | EIASR 99 Part II |
| **Mineral extraction** | Disposal of mineral waste<br>Exploratory deep drilling<br>Extraction of minerals at the surface by open casting/quarrying<br>Extraction of minerals by underground mining<br>Fluvial dredging<br>Peat extraction (commercial)<br>Installations for the processing of specified minerals/products | EIASR 99 Part II |
| | Marine dredging | |
| **Transport and communications** | Review of old mineral permissions | EIASROMPRO2 |
| | Docks, harbours, ports, piers and jetties and ferry terminals | HarbourWEIAR 99 |
| | Airfields, airports, runways<br>Inland waterways and canals/canalisation for transport<br>Intermodal trans-shipment facilities and terminals<br>Light railways and tram systems<br>Motorway service areas<br>Railways | EIASR 99 Part II |
| | Pipelines to carry chemicals | PipelineWEIAR 00 |
| | Roads | EIASR 99 Part III |

| Sector | Project type | EIA Regulations |
|---|---|---|
| **Urban developments** | Business parks, industrial estates and employment developments<br>Housing estates<br>New settlements<br>Retail parks and other retail developments | EIASR 99 Part II |
| **Waste management** | Deposit of dredgings on land<br>Disposal of mineral waste<br>Disposal of hazardous wastes<br>Incinerators and other installations for waste disposal<br>Landfill and land-raise<br>Scrap yards<br>Sludge deposition<br>Waste water treatment plants and outfalls | EIASR 99 Part II |
| | Deposit of dredgings at sea | |
| **Water** | Dams and installations designed to hold or store water<br>Development for abstraction from river systems<br>Development for artificial recharge systems<br>Development for abstraction from ground waters<br>Development for water treatment and supply<br>Development for transfer of water between river basins<br>Long distance aqueducts | EIASR 99 Part II |
| | Abstraction of water, irrigation, drainage and other water management projects for agriculture | EIAWaterMR03 |

## The following abbreviations of the EIA Regulations are used in Annexe 2 Tables 2 and 3

| | |
|---|---|
| EIAFishFarmMWR 99 | Environmental Impact Assessment (Fish Farming in Marine Waters) Regulations 1999 |
| EIAForestrySR 99 | Environmental Impact Assessment (Forestry) (Scotland) Regulations 1999 |
| EIASR 99 | Environmental Impact Assessment (Scotland) Regulations 1999 |
| EIASROMPR02 | The Environmental Impact Assessment (Scotland) Amendment Regulations 2002 (Review of Old Mineral Permissions (ROMPs)) |
| EIAWaterMR03 | The Environmental Impact Assessment (Water Management) (Scotland) Regulations 2003 |
| ElecWorks EIASR 00 | Electricity Works (Environmental Impact Assessment) (Scotland) Regulations 2000 |
| HarbourWEIAR 99 | Harbour Works (Environmental Impact Assessment) Regulations 1999 |
| NuclearREIADR 99 | Nuclear Reactors (Environmental Impact Assessment for Decommissioning) Regulations 1999 |
| OffshorePPPAEER 99 | Offshore Petroleum Production and Pipelines (Assessment of Environmental Effects) Regulations 1999 |
| PGasTransPWEIAR 99 | Public Gas Transporters (Pipeline Works) (Environmental Impact Assessment) Regulations 1999 |
| PipelineWEIAR 00 | Pipeline Works (Environmental Impact Assessment) Regulations 2000 |
| OffshoreGenStnsR02 | The Electricity Act 1989 (Requirement of Consent for Offshore Generating Stations) (Scotland) Order 2002 SSI 2002 No. 407 |
| ULSNAR02 | The Environmental Impact Assessment (Uncultivated Land and Semi-Natural Areas) (Scotland) Regulations 2002 |

## Key to Annexe 2 Table 3 below

| | |
|---|---|
| Interpretation | Interpretation, including definitions |
| Compliance | The requirement to comply with the regulations before granting consents |
| Screening | Screening to establish whether EIA will apply |
| Scoping | Scoping of the environmental statement |
| Application without ES | What happens where an application is made without an environmental statement |
| SNH to give info | The provisions requiring SNH to give information to help the proposer compile the statement |
| Publicity | Provisions for publicity |
| Consultations | Requirements for consultations |
| Further info | The powers to require further information or evidence to be submitted |
| Transboundary | Provisions for dealing with potential transboundary effects affecting another EC member state |
| Public Cons period | The statutory minimum public consultation/notification period |
| SNH Cons period | The statutory minimum period allowed for SNH to reply to a consultation (if specified) |
| Final decision/records | Requirements for making and recording the Competent Authority's decision |
| Schedule 1 projects | The definition of Schedule 1 projects |
| Schedule 2 projects | The definition of Schedule 2 projects |
| Matters to consider | The matters to be considered if determining whether a project is EIA development subject to the EIA procedure |
| Content of ES | The requirements for the content of environmental statements |

**Annexe 2 Table 3 Key Information and References in EIA Regulations    Part 1**

| Project Type | Devlpmt requiring PP | Develpmt by a PA inc local roads | Unauthorised develpmt appeal | Review of Old Mineral Permissions | Motorways and trunk roads | Drainage Improvements | Marine Aquaculture |
|---|---|---|---|---|---|---|---|
| **Competent Authority** | PA or Scottish Ministers | PA or Scottish Ministers | Scottish Ministers | PA or Scottish Ministers | Scottish Ministers | Scottish Ministers | CEC, Orkney Islands Council and Shetland CC |
| **Consent Procedure** | Planning permission under TCP (Scotland) Act 1997 | Notice of intention to develop under TCP (Develpmt by PA's) (Scotland) Regulations 1981 | Enforcement Notice appeal under TCP (Scotland) Act 1997 | Review process of Mineral Permissions granted between 1948 and 1982 and all later permissions every 15 yrs | SM decide to proceed or make order under Sch. 1 Roads (Scotland) Act 1980 | SM consent under Land Drainage (Scotland) Act 1958 | CEC consent for fish farming in marine waters + licence from OIC or SCC |
| **EIA Regulations** | EIASR 99 | EIASR 99 | EIASR 99 | EIAS(Amendment) (ROMPs) Regs 02 | EIASR 99 | EIASR 99 | EIAFishFarmMWR 99 |
| **Jurisdiction** | Scotland | Scotland | Scotland | Scotland | Scotland | Scotland | Great Britain |
| **Statutory Instrument** | SSI 1999/1 | SSI 1999/1 | SSI 1999/1 | SSI 2002/324 | SSI 1999/1 | SSI 1999/1 | SI1999/367 |
| **Came into force** | 1.8.1999 | 1.8.1999 | 1.8.1999 | 23.9.2002 | 1.8.1999 | 1.8.1999 | 14.3.1999 |
| **Interpretation** | Reg. 2 | Reg. 2 | Reg. 2 | Reg. 2 of 1999 Regs as amended | Reg. 2 | Regs 2 & Reg 55 | Reg. 2 |
| **Compliance** | Reg. 3 | Regs 22–25 | Reg. 29 | Regs 3 and 28A of 1999 Regs as amended | Regs 49 & 50 amend S20A & 55A RSA 1980 | Reg. 57 | Reg. 3 |
| **Screening** | Reg. 4–6 | Reg. 22 | Regs 30–31 | Regs 4–6 and 28A of 1999 Regs as amended | N/A | Reg. 56 | Reg. 4 |
| **Scoping** | Regs 10–11 | N/A | N/A | Regs 10–11 and 28A of 1999 Regs as amended | N/A | N/A | Reg. 6 |

| | | | | | | | |
|---|---|---|---|---|---|---|---|
| **Application without ES** | Regs 7 + 9 | Reg. 22 | Regs 33–34 | Regs 7–9 and 28A of 1999 Regs as amended | N/A | N/A | Reg. 5 |
| **SNH to give info.** | Reg. 12 | Regs 22–23 | Reg. 32 | Reg. 12 of 1999 Regs as amended | N/A | Reg. 58 | Reg. 7 |
| **Publicity** | Regs 13–18 | Reg. 24 | Regs 37–38 | Regs 13–18 and 28A of 1999 Regs as amended | Regs 49 & 50 amend S20A & 55A RSA 80 | Reg. 59 | Reg. 8 |
| **Consultations** | Regs 14 + 16 | Reg. 24(3) | Reg. 35 | Regs 14 + 16 of 1999 Regs as amended | | Reg. 59 | Reg. 9 |
| **Further info** | Reg. 19 | Reg. 24(4) | Reg 36 | Regs 19 and 28A of 1999 Regs as amended | N/A | Reg. 60 | Reg. 10 |
| **Transboundary** | Regs 40–41 | Regs 40–41 | Regs 39–41 | Regs 39–41 of 1999 Regs as amended | Reg. 49 amends S20B & 55B RSA 80 | N/A | N/A |
| **Public Cons period** | 4 weeks | 4 weeks after publicity | 2 weeks after a min 3 wk publicity | 4 weeks | 3 weeks | 28 days (LDSA 58) | 28 days |
| **SNH Cons period** | 4 weeks | Unspecified | Unspecified | 4 weeks | Opportunity to express an opinion | 28 days (LDSA 58) | 28 days |
| **Final decision/ records** | Regs 20–21 | Reg. 26 | Reg. 38 | Regs 20–21 of 1999 Regs as amended | Reg. 52 amends Sch. 1 RSA 80 | N/A | Reg. 11 |
| **Schedule 1 projects** | Sch. 1 | Sch. 1 | Sch. 1 | Sch. 1 of 1999 Regs | Sch. 1 | N/A | N/A |
| **Schedule 2 projects** | Sch. 2 | Sch. 2 | Sch. 2 | Sch. 2 of 1999 Regs | Sch. 2 | Sch. 2 | N/A |
| **Matters to consider** | Sch. 3 | Sch. 3 | Sch. 3 | Sch. 3 of 1999 Regs | Sch. 3 | Sch. 3 | N/A |
| **Content of ES** | Sch. 4 | Sch. 4 | Sch. 4 | Sch. 4 of 1999 Regs | Sch. 4 | Sch. 4 | N/A |

**Annexe 2 Table 3 Key Information and References in EIA Regulations   Part 2**

| Project Type | Forestry Works | Uncultivated or Semi-Natural Areas | Water Management for Agriculture | Electricity power stations (over 50MW) and overhead lines | Offshore Electricity power stations over 1MW | Gas Pipelines not requiring PP | Offshore Oil and Gas and Pipelines |
|---|---|---|---|---|---|---|---|
| Competent Authority | Forestry Commissioners or on appeal Scottish Ministers | Scottish Ministers (SEERAD) | PA or Scottish Ministers | Scottish Ministers | Scottish Ministers | Scottish Ministers | Secretary of State |
| Consent Procedure | Consent of the FC for afforestation, deforestation, forest tracks and quarries | All projects to be screened, relevant projects require consent under the Regs | Relevant irrigation or drainage or other water management works for agriculture require planning permission | All power stations on and offshore over 50MW require SM consent under S.36 (Power Station) or S.37 (Overhead lines) of Electricity Act 1989 | All offshore power stations driven mainly by water or wind over 1MW require consent of SM under S.36 of Electricity Act 1989 so Electricity Works EIASR 00 apply | Consent under Reg. 14 of these Regs for develpmt under Part 17 Class Fa of the TCP (GPDO) 92 where subject to an environmental determination by SM | Prior consent required by a licence to explore, produce or transport oil & gas granted under provisions of Petroleum Act 1998 |
| EIA Regulations | EIAForestySR 99 | EIA(ULSNA)Regs 2002 | EIAWaterMR03 | Electricity Works EIASR 00 | Offshore GenStns Regs 02 | PGasTransPWEIAR 99 | OffshorePetrolPPAE ER 99 |
| Jurisdiction | Scotland | Scotland | Scotland | Scotland | Scotland | Great Britain | UK |
| Statutory Instrument | SSI 1999/43 | SSI 2002/6 | SSI 2003/341 | SSI 2000/320 | SSI 2002/407 | SI 1999/1672 | SI 1999/360 |
| Came into force | 6.9.1999 | 4.2.2002 | 30.9.2003 | 5.10.2000 | 26.9.2002 | 15.7.1999 | 14.3.1999 |
| Interpretation | Reg. 2, 3 & 15 | Reg. 2 | Reg. 2 EIASR99 | Reg. 2 | Reg. 2 of 2002 Regs | Reg. 2 | Reg. 3 |
| Compliance | Reg. 4 | Regs 3, 4 and 6 | Reg. 3 EIASR99 | Regs 3-4 | Regs 3-4 of 2002 Regs | Reg. 3 | Regs 4, 5 + 11 |
| Screening | Regs 5-8 | Regs 4 and 5 | Reg. 4-6 EIASR99 | Reg. 5 | Reg. 5 of 2002 Regs | Reg. 6 | Regs 6, 11 + 12 |
| Scoping | Reg. 9 | Regs 6 and 7 | Regs 10-11 EIASR99 | Reg. 7 | Reg. 7 of 2002 Regs | Reg. 7 | Reg. 7 |

| | | | | | | | |
|---|---|---|---|---|---|---|---|
| **Application without ES** | N/A | Reg. 9 | Regs 7 + 9 EIASR99 | Reg. 6 | Reg. 6 of 2002 Regs | Reg. 9 | Reg. 5 |
| **SNH to give info.** | Reg. 12 | Reg. 8 | Reg. 12 EIASR99 | Regs 8 + 15 | Reg. 8 + 15 of 2002 Regs | Reg. 8 | Regs 5 + 8 |
| **Publicity** | Reg. 13 | Reg. 9(2) | Regs 13–18 EIASR99 | Regs 9–11 + 14 | Reg. 9–11 + 14 of 2002 Regs | Reg. 10 | Regs 9 + 10 |
| **Consultations** | Regs 20–23 (enforce) | Reg. 9(2) | Regs 14 + 16 EIASR99 | Reg. 11 | Reg. 11 of 2002 Regs | Reg. 10 | Regs 9 + 10 |
| **Further info** | Reg. 11 | Reg. 10 | Reg. 19 EIASR99 | Reg. 13 | Reg. 13 of 2002 Regs | Reg. 11 | Reg. 10 |
| **Transboundary** | Reg. 14 | Regs 11 and 12 | Regs 40–41 EIASR99 | Reg. 12 | Reg. 12 of 2002 Regs | Reg. 13 | Regs 5 + 12 |
| **Public Cons period** | 28 days | 42 days (28 for further information) | 4 weeks | 4 weeks after publicity | 4 weeks after publicity | 28 days | 4 weeks |
| **SNH Cons period** | 28 days | 42 days | 4 weeks | 14 days from receipt of ES | 14 days from receipt of ES | 28 days | 4 weeks |
| **Final decision / records** | Regs 15, 16 + 24 | Regs 13 and 14 | Regs 20–21 EIASR99 | N/A | N/A | Reg. 8 | Reg. 5 |
| **Schedule 1 projects** | N/A | N/A | Sch. 1 EIASR99 | Sch. 1 | Sch. 1 of 2002 Regs | Sch. 3 | N/A |
| **Schedule 2 projects** | N/A | N/A | Sch. 2 EIASR99 as amended | Sch. 2 | Sch. 2 of 2002 Regs | Sch. 3 | N/A |
| **Matters to consider** | Sch. 2 & 3 | Sch. 1 | Sch. 3 EIASR99 | Sch. 3 | Sch. 3 of 2002 Regs | Sch. 2 | Sch. 1 |
| **Content of ES** | Sch. 1 | Sch. 2 | Sch. 4 EIASR99 | Sch. 4 | Sch. 4 of 2002 Regs | Sch. 1 | Sch. 2 |

**Annexe 2 Table 3 Key Information and References in EIA Regulations   Part 3**

| Project Type | Other Pipelines | Decommissioning Nuclear Installations | Harbours, Docks, Piers and Ferries |
|---|---|---|---|
| **Competent Authority** | Secretary of State | Health & Safety Executive | Scottish Ministers |
| **Consent Procedure** | Pipeline construction authorisation under Pipelines Act 1962 | Nuclear Installations Act 1965 Licensees apply for consent under Reg. 8 | Consents under S.34 or S.35 Coast Protection Act 1949; S.37 Merchant Shipping Act 1988; any local Act; Harbour Revision or Empowerment Orders under Harbours Act 1964 |
| **EIA Regs** | PipelineWEIAR 00 | NuclearREIADR 99 | HarbourWEIAR 99 |
| **Jurisdiction** | E, S & W | Great Britain | E, S & W |
| **Statutory Instrument** | SI 2000/1928 | SI 1999/2892 | SI 1999/3445 amended by SI 2000/2391 |
| **Came into force** | 1.9.2000 | 19.11.1999 | 1.2.1999 |
| **Interpretation** | Reg. 2 | Reg. 2 | Reg. 2 & Sch. 3(1) HA 64 |
| **Compliance** | Reg. 3 | Regs 3–5 + 8 | Regs 5 + 6 & Sch. 3(3–6) HA 64 |
| **Screening** | Reg. 4 | N/A | Reg. 4 & Sch. 3(5) HA 64 |
| **Scoping** | Reg. 5 | Reg. 6 | Reg. 4 & Sch. 3(6) HA 64 |
| **Application without ES** | Regs 11–13 | Reg. 16 | Regs 11–14 |
| **SNH to give info.** | Reg. 6 | Reg. 7 | |
| **Publicity** | Reg. 7–8 | Reg. 9 | Regs 7 & 9 & Sch. 3(14–15) HA 64 |
| **Consultations** | Reg. 7 | Regs 8–9 | |
| **Further info** | Reg. 8 | Reg. 10 | |
| **Transboundary** | Reg. 3 | Regs 8 + 12 | Reg. 8 & Sch. 3(16) HA 64 |
| **Public Cons period** | 28 days | 30 days | 42 days |
| **SNH Cons period** | 28 days | Such reasonable time as HSE may specify + 14 days for further info | Reasonable opportunity Sch. 3 16(5) HA 64 |

| | | | |
|---|---|---|---|
| **Final decision/records** | Reg. 3 | Reg. 11 | Reg. 10 & Sch. 3(19–25) HA 64 |
| **Schedule 1 projects** | N/A | N/A | N/A |
| **Schedule 2 projects** | N/A | N/A | N/A |
| **Matters to consider** | Sch. 2 | Sch. 2 | Sch. 2 |
| **Content of ES** | Sch. 1 | Sch. 1 | Sch. 1 & Sch. 3(8) HA 64 |

## Other abbreviations used in Annexe 2 Table 3

| | |
|---|---|
| CEC | Crown Estate Commissioners |
| Devlpmt | Development |
| E, S & W | England, Scotland and Wales |
| GB | Great Britain |
| HA 64 | Harbours Act 1964 |
| LDSA 58 | Land Drainage (Scotland) Act 1958 |
| PA | Planning Authority |
| PP | Planning Permission |
| Reg. | Regulation |
| RSA 80 | Roads (Scotland) Act 1980 |
| S. | Section (of Act) |
| Sch. | Schedule |
| SM | Scottish Ministers |
| TCP | Town and Country Planning |
| UK | United Kingdom inc. territorial waters |
| DPASR 81 | Town and Country Planning (Development by Planning Authorities) (Scotland) Regs 1981 |

# Annexe 3 List of Current Relevant National Policy and Guidance

| Guidance | Commentary |
|---|---|
| Scottish Office Circular 26/1991 *Environmental Assessment and Private Legislation Procedures* | This Circular provides administrative, procedural and policy guidance on the Government's procedures for ensuring compliance with the Environmental Assessment directive in respect of projects to be authorised directly by Parliament because they may not be subject to normal planning or other consenting procedures that would include Environmental Assessment where necessary. The Appendices are amended in line with Circular 15/1999. |
| Scottish Office Circular 3/1991 *Electricity Generating Stations and Overhead Lines: Permitted Development for Electricity Undertakings* | This Circular provides administrative, procedural and policy guidance on projects for or at power stations and for overhead lines that may otherwise be permitted development and for which no planning application would therefore be made. |
| Scottish Executive Development Department PAN 51, 1997, *Planning and Environmental Protection* | This Planning Advice Note provides background information and advice on good practice in the planning process with reference to pollution control and other forms of environmental protection, with obvious relevance to the EIA process. |
| Scottish Executive Circular June 2000 *Habitats and Birds Directives Nature Conservation: Implementation in Scotland of the EC Directives on the Conservation of Natural Habitats and of Wild Flora and Fauna and the Conservation of Wild Birds*. Amends Scottish Office Circular 6/1995 Habitats and Birds Directives *The Conservation (Natural Habitats Etc.) Regulations 1994*. | Provides procedural and policy guidance on the Habitats Regulations 1994, and specifically indicates that any project likely to have a significant effect on a Natura 2000 (European) Site, whether fully designated or not, should normally be subject to the Environmental Impact Assessment process. The Circular also explains how this differs from the appropriate assessment undertaken by the Competent Authority under the Habitats Regulations. |
| Scottish Executive Development Department Circular 15/1999 *The Environmental Impact Assessment (Scotland) Regulations 1999* | This Circular provides comprehensive guidance on the EIA process with particular emphasis on projects requiring planning permission and those requiring approval under the Roads (Scotland) Act or the Land Drainage Acts. |
| Scottish Executive Development Department PAN 58, 1999, *Environmental Impact Assessment* | This Planning Advice Note provides background information and advice on good practice in the EIA process to supplement the legal, administrative and policy advice in Circular 15/1999. |

SEERAD Guidelines on Environmental Impact Assessment (EIA) for use of uncultivated land and semi-natural areas for intensive agricultural purposes Feb 2002

Provides guidance on the ULSNA Regulations including the new consenting and screening procedures

Environmental Impact Assessment (EIA) Directive 1] Minimum Requirements of the Regulations and 2] Outline Planning Applications, letter to all Heads of Planning, from SEDD Planning Division, June 2002

Together with a note attached to the letter with questions and answers, provides guidance to planning authorities on minimising the risk of legal challenges and how to deal with outline planning applications

SEDD Circular 1/2003 *The Environmental Impact Assessment (Scotland) Regulations 2002 Review of Old Mineral Permissions (ROMPs)*, Jan 2003

Explains the regulations that introduced the requirement to apply EIA procedures to the review of old mineral permissions in order to comply with a court ruling that the review and issue of new conditions amounts to the grant of a new consent that should be subject to EIA.

SEDD Circular 3/2003 *The Environmental Impact Assessment (Water Management) (Scotland) Regulations 2003*, Nov 2003

Explains the regulations that amend the definition of development to include carrying out of irrigation or drainage or other water management works for agriculture so making such projects potentially EIA development subject to the EIASR 99.

# Annexe 4   Projects Requiring Environmental Impact Assessment

## Schedule 1 Developments requiring EIA in every case

1) Crude oil refineries (excluding undertakings manufacturing only lubricants from crude oil) and installations for the gasification and liquefaction of 500 tonnes or more of coal or bituminous shale per day.

2) Thermal power stations and other combustion installations with a heat output of 300 megawatts or more and nuclear power stations and other nuclear reactors (except research installations for the production and conversion of fissionable and fertile materials whose maximum power does not exceed 1 kilowatt continuous thermal load).

3) Installations for the reprocessing of irradiated nuclear fuel; installations designed for the production or enrichment of nuclear fuel; for the processing of irradiated nuclear fuel or high-level radioactive waste; for the final disposal of irradiated nuclear fuel; solely for the final disposal of radioactive waste; solely for the storage (planned for more than 10 years) of irradiated nuclear fuels or radioactive waste in a different site than the production site.

4) Integrated works for the initial smelting of cast iron and steel. Installations for the production of non-ferrous crude metals (as described and further specified in Schedule 1(4) of the EIASR 99).

5) Installations for the extraction of asbestos and for the processing and transformation of asbestos and products containing asbestos

   a. where the installation produces asbestos-cement products, with an annual production of more than 20,000 tonnes of finished products,
   b. where the installation produces friction material, with an annual production of more than 50 tonnes of finished products, and
   c. other cases where the installation will utilise more than 200 tonnes of asbestos per year.

6) Integrated chemical installations (as described and further specified in Schedule 1(6) of the EIASR 99).

7) Construction of motorways, express roads and other roads of four or more lanes and the realignment or widening of roads to provide four or more lanes where the road would be 10 km or more continuous length. Lines for long-distance railway traffic and airports with a basic runway length of 2100 m or more.

8) Trading ports and construction of piers for loading and unloading connected to land outside ports and also inland waterways and ports for inland waterway traffic which permit the passage of vessels of over 1350 tonnes.

9) Waste-disposal installations for the incineration or chemical treatment or landfill of hazardous waste.

10)    Incineration or chemical treatment of non-hazardous wastes (installations with a capacity of more than 100 tonnes per day).

11)    Ground water abstraction or artificial recharge schemes exceeding 10 million m$^3$ per year.

12)    Transfer of water resources other than piped drinking water between river basins above 100 million m$^3$ per year or over 5% of flows where the abstracted river exceeds a flow of 2000 million m$^3$ per year.

13)    Waste water treatment plants (over 150,000 population equivalents).

14)    Extraction of petroleum (more than 500 tonnes per day) and natural gas (over 500,000 m$^3$ per day).

15)    Dams and similar installations, with water holdback capacity exceeding 10 million m.

16)    Pipelines to transport oil, gas or chemicals (more than 40 km long and 800 mm diameter).

17)    Installations for intensive rearing of poultry or pigs above 85,000 broilers, 60,000 hens, 3,000 pigs over 30 kg or 900 sows.

18)    All pulp and those paper and board factories over 200 tonnes per day production.

19)    Quarries and opencast mining (over 25 ha) and peat extraction (over 150 ha).

20)    Installations for storage of petrol and petrochemical products (200,000 tonnes and over).

## Schedule 2 Developments

Requiring assessment if they are likely to have significant effects on the environment by virtue, inter alia, of their nature size or location. The carrying out of development to provide any of the following:

| Description of development | Applicable thresholds/criteria |
|---|---|
| **1. Agriculture and aquaculture** | |
| 1(a) Projects for the use of uncultivated land or semi-natural areas for intensive agricultural purposes | The area of the development exceeds 0.5 hectare |
| 1(b) Water management projects for agriculture, including irrigation and land drainage projects | The area of the works exceeds 1 hectare |
| 1(c) Intensive livestock installations (unless included in Schedule 1) | The area of new floorspace exceeds 500 square metres |
| 1(d) Intensive fish farming | The installation resulting from the development is designed to produce more than 10 tonnes of dead weight fish per year |
| 1(e) Reclamation of land from the sea | All development |
| **2. Extractive industry** | |
| 2(a) Quarries, open-cast mining and peat extraction (unless included in Schedule 1) | All development except the construction of buildings or other ancillary structures where the new floorspace does not exceed 1000 square metres |
| 2(b) Underground mining | |
| 2(c) Extraction of minerals by marine or fluvial dredging | All development |
| 2(d) Deep drillings, in particular: (i) geothermal drilling; (ii) drilling for the storage of nuclear waste material; (iii) drilling for water supplies; with the exception of drillings for investigating the stability of the soil. | (i) In relation to any type of drilling, the area of the works exceeds 1 hectare; or (ii) in relation to geothermal drilling and drilling for the storage of nuclear waste material, the drilling is within 100 metres of any controlled waste |
| 2(e) Surface industrial installations for the extraction of coal, petroleum, natural gas and ores, as well as bituminous shale | The area of the development exceeds 0.5 hectare |
| **3. Energy industry** | |
| 3(a) Industrial installations for the production of electricity, steam and hot water (unless included in Schedule 1) | The area of the development exceeds 0.5 hectare |
| 3(b) Industrial installations for carrying gas, steam and hot water | The area of the works exceeds 1 hectare |

| Description of development | Applicable thresholds/criteria |
|---|---|
| 3(c) Surface storage of natural gas<br>3(d) Underground storage of combustible gases<br>3(e) Surface storage of fossil fuels | (i) The area of any new building, deposit or structure exceeds 500 square metres; or<br>(ii) a new building, deposit or structure is to be sited within 100 metres of any controlled waters |
| 3(f) Industrial briquetting of coal and lignite | The area of new floorspace exceeds 1000 square metres |
| 3(g) Installations for the processing and storage of radioactive waste (unless included in Schedule 1) | (i) The area of new floorspace exceeds 1000 square metres; or<br>(ii) the installation resulting from the development will require an authorisation or the variation of an authorisation under the Radioactive Substances Act 1993 |
| 3(h) Installations for hydroelectric energy production | The installation is designed to produce more than 0.5 megawatts |
| 3(i) Installations for the harnessing of wind power for energy production (wind farms) | (i) The development involves the installation of more than two turbines; or<br>(ii) the hub height of any turbine or height of any other structure exceeds 15 metres |

## 4. Production and processing of metals

| | |
|---|---|
| 4(a) Installations for the production of pig iron or steel (primary or secondary fusion) including continuous casting<br>4(b) Installations for the processing of ferrous metals:<br>(i) hot-rolling mills;<br>(ii) smitheries with hammers;<br>(iii) application of protective fused metal coats<br>4(c) Ferrous metal foundries<br>4(d) Installations for the smelting, including the alloyage, of non-ferrous metals, excluding precious metals, including recovered products (refining, foundry casting, etc.)<br>4(e) Installations for surface treatment of metals and plastic materials using an electrolytic or chemical process<br>4(f) Manufacture and assembly of motor vehicles and manufacture of motor-vehicle engines<br>4(g) Shipyards<br>4(h) Installations for the construction and repair of aircraft<br>4(i) Manufacture of railway equipment<br>4(j) Swaging by explosives<br>4(k) Installations for the roasting and sintering of metallic ores | The area of new floorspace exceeds 1000 square metres |

| Description of development | Applicable thresholds/criteria |
|---|---|
| **5. Mineral industry** | |
| 5(a) Coke ovens (dry coal distillation) <br> 5(b) Installations for the manufacture of cement <br> 5(c) Installations for the production of asbestos and the manufacture of asbestos-based products (unless included in Schedule 1) <br> 5(d) Installations for the manufacture of glass including glass fibre <br> 5(e) Installations for smelting mineral substances including the production of mineral fibres <br> 5(f) Manufacture of ceramic products by burning, in particular roofing tiles, bricks, refractory bricks, tiles, stoneware or porcelain | The area of new floorspace exceeds 1000 square metres |
| **6. Chemical industry (unless included in Schedule 1)** | |
| 6(a) Treatment of intermediate products and production of chemicals <br> 6(b) Production of pesticides and pharmaceutical products, paint and varnishes, elastomers and peroxides | The area of new floorspace exceeds 1000 square metres |
| 6(c) Storage facilities for petroleum, petrochemical and chemical products | (i) The area of any new building or structure exceeds 0.5 hectare; or <br> (ii) more than 200 tonnes of petroleum, petrochemical or chemical products is to be stored at any one time. |
| **7. Food industry** | |
| 7(a) Manufacture of vegetable and animal oils and fats <br> 7(b) Packing and canning of animal and vegetable products <br> 7(c) Manufacture of dairy products <br> 7(d) Brewing and malting <br> 7(e) Confectionery and syrup manufacture <br> 7(f) Installations for the slaughter of animals <br> 7(g) Industrial starch manufacturing installations <br> 7(h) Fish meal and fish oil factories <br> 7(i) Sugar factories | The area of new floorspace exceeds 1000 square metres |
| **8. Textile, leather, wood and paper industries** | |
| 8(a) Industrial plants for the production of paper and board (unless included in Schedule 1) <br> 8(b) Plants for the pre-treatment (operations such as washing, bleaching, mercerisation) or dyeing of fibres or textiles <br> 8(c) Plants for the tanning of hides and skins <br> 8(d) Cellulose processing and production installations. | The area of new floorspace exceeds 1000 square metres |

| Description of development | Applicable thresholds/criteria |
|---|---|
| **9. Rubber industry** | |
| Manufacturing and treatment of elastomer-based products. | The area of new floorspace exceeds 1000 square metres |
| **10. Infrastructure projects** | |
| 10(a) Industrial estate development projects<br>10(b) Urban development projects, including the construction of shopping centres and car parks, sports stadiums, leisure centres and multiplex cinemas<br>10(c) Construction of intermodal transshipment facilities and of intermodal terminals (unless included in Schedule 1) | The area of the development exceeds 0.5 hectare |
| 10(d) Construction of railways (unless included in Schedule 1) | The area of the works exceeds 1 hectare |
| 10(e) Construction of airfields (unless included in Schedule 1) | (i) The development involves an extension to a runway; or<br>(ii) the area of the works exceeds 1 hectare |
| 10(f) Construction of roads (unless included in Schedule 1) | The area of the works exceeds 1 hectare |
| 10(g) Construction of harbours and port installations, including fishing harbours (unless included in Schedule 1) | The area of the works exceeds 1 hectare |
| 10(h) Inland waterway construction not included in Schedule 1, canalisation and floor-relief works<br>10(i) Dams and other installations designed to hold water or store it on a long-term basis (unless included in Schedule 1)<br>10(j) Tramways, elevated and underground railways, suspended lines or similar lines of a particular type, used exclusively or mainly for passenger transport | The area of the works exceeds 1 hectare |
| 10(k) Oil and gas pipeline installations (unless included in Schedule 1)<br>10(l) Installations of long-distance aquaducts | (i) The area of the works exceeds 1 hectare; or<br>(ii) in the case of a gas pipeline, the installation has a design operating pressure exceeding 7 bar gauge |
| 10(m) Coastal work to combat erosion and maritime works capable of altering the coast through the construction, for example, of dykes, moles, jetties and other sea defence works, excluding the maintenance and reconstruction of such works | All development |

| Description of development | Applicable thresholds/criteria |
|---|---|
| 10(n) Groundwater abstraction and artificial groundwater recharge schemes not included in Schedule 1<br>10(o) Works for the transfer of water resources between river basins not included in Schedule 1 | The area of the works exceeds 1 hectare |
| 10(p) Motorway service areas | The area of the development exceeds 0.5 hectare |

## 11. Other projects

| | |
|---|---|
| 11(a) Permanent racing and test tracks for motorised vehicles | The area of the development exceeds 1 hectare |
| 11(b) Installations for the disposal of waste (unless included in Schedule 1) | (i) The disposal is by incineration; or<br>(ii) the area of the development exceeds 0.5 hectare; or<br>(iii) the installation is to be sited within 100 metres of any controlled waters |
| 11(c) Waste-water treatment plants (unless included in Schedule 1) | The area of the development exceeds 1000 square metres |
| 11(d) Sludge-deposition sites<br>11(e) Storage of scrap iron, including scrap vehicles | (i) The area of deposit or storage exceeds 0.5 hectare; or<br>(ii) a deposit is to be made or scrap stored within 100 metres of any controlled waters |
| 11(f) Test benches for engines, turbines or reactors<br>11(g) Installations for the manufacture of artificial mineral fibres<br>11(h) Installations for the recovery or destruction of explosive substances<br>11(i) Knackers' yards | The area of new floorspace exceeds 1000 square metres. |

## 12. Tourism and leisure

| | |
|---|---|
| 12(a) Ski-runs, ski-lifts and cable cars and associated developments | (i) The area of the works exceeds 1 hectare; or<br>(ii) the height of any building or other structure exceeds 15 metres |
| 12(b) Marinas | The area of the enclosed water surface exceeds 1000 square metres |
| 12(c) Holiday villages and hotel complexes outside urban areas and associated developments<br>12(d) Theme parks | The area of the development exceeds 0.5 hectare |
| 12(e) Permanent camp sites and caravan sites<br>12(f) Golf courses and associated developments. | The area of the development exceeds 1 hectare |

| Description of development | Applicable thresholds/criteria |
|---|---|
| **13. Changes or extensions** | |
| 13(a) Any change to or extension of development of a description listed in Schedule 1 or in paragraphs 1 to 12 of Column 1 of this table, where that development is already authorised, executed or in the process of being executed, and the change or extension may have significant adverse effects on the environment<br><br>Any change to or extension of development of a description listed in Schedule 1 or in paragraphs 1 to 12 of Column 1 of this table, where that development is already authorised, executed or in the process of being executed, and the change or extension may have significant adverse effects on the environment | (i) In relation to development of a description mentioned in Column 1 of this table, the thresholds and criteria in the corresponding part of Column 2 of this table applied to the change or extension (and not to the development as changed or extended).<br>(ii) In relation to development of a description mentioned in a paragraph in Schedule 1 indicated below, the thresholds and criteria in Column 2 of the paragraph of this table indicated below applied to the change or extension (and not to the development as changed or extended: |

| Paragraph in Schedule 1 | Paragraph of this table |
|---|---|
| 1 | 6(a) |
| 2(a) | 3(a) |
| 2(b) | 3(g) |
| 3 | 3(g) |
| 4 | 4 |
| 5 | 5 |
| 6 | 6(a) |
| 7(a) | 10(d) (in relation to railways) or 10(e) (in relation to airports) |
| 7(b) and (c) | 10(f) |
| 8(a) | 10(h) |
| 8(b) | 10(g) |
| 9 | 11(b) |
| 10 | 11(b) |
| 11 | 10(n) |
| 12 | 10(o) |
| 13 | 11(c) |
| 14 | 2(e) |
| 15 | 10(i) |
| 16 | 10(k) |
| 17 | 1(c) |
| 18 | 8(a) |
| 19 | 2(a) |
| 20 | 6(c) |

| Description of development | Applicable thresholds/criteria |
|---|---|
| 13(b) Development of a description mentioned in Schedule 1, undertaken exclusively or mainly for the development and testing of new methods or products and used for not more than 2 years | All development |

# Indicative Thresholds and Criteria for Identification of Schedule 2 Development Requiring EIA

(Annexe A Circular 15/1999)

The criteria and thresholds in this Annexe are only indicative and reference must be made to Section C.1 of the handbook when using this Annexe.

| Class of project | Project type | Threshold/criteria |
|---|---|---|
| Agricultural Development | General | In general, agricultural operations fall outside the scope of the Town and Country Planning system and, where relevant, will be regulated under other consent procedures. The descriptions below apply only to projects that are considered to be 'development' for the purposes of the Town and Country Planning (Scotland) Act 1997. |
| | Use of uncultivated or semi-natural land for intensive agricultural purposes | Development (such as greenhouses, farm buildings etc.) on previously uncultivated land is unlikely to require EIA unless it covers more than 5 hectares. In considering whether particular development is likely to have significant effects, consideration should be given to impacts on the surrounding ecology, hydrology and landscape. |
| | Water management for agriculture, including irrigation and land drainage works | EIA is more likely to be required if the development would result in permanent changes to the character of more than 5 hectares of land. In assessing the significance of any likely effects, particular regard should be had to whether the development would have damaging wider impacts on hydrology and surrounding ecosystems. It follows that EIA will not normally be required for routine water management projects undertaken by farmers. |
| | Intensive livestock installations | The significance or otherwise of the impacts of intensive livestock installations will often depend upon the level of odours, increased traffic and the arrangements for waste handling. EIA is more likely to be required for intensive livestock installations if they are designed to house more than 750 sows, 2000 fattening pigs, 60,000 broilers or 50,000 layers, turkeys or other poultry. |
| | Intensive fish farming | Apart from the physical scale of any development, the likelihood of significant effects will generally depend on the extent of any likely wider impacts on the hydrology and ecology of the surrounding area. Developments designed to produce more than 100 tonnes (dead weight) of fish per year will be more likely to require EIA. |
| | Reclamation of land from the sea | In assessing the significance of any development, regard should be had to the likely wider impacts on natural coastal processes beyond the site itself, as well as to the scale of reclamation works themselves. EIA is more likely to be required where work is proposed on a site which exceeds 1 hectare. |

| Class of project | Project type | Threshold/criteria |
|---|---|---|
| Extractive Industry | Surface and underground mineral working | The likelihood of significant effects will tend to depend on the scale and duration of the works, and the likely consequent impact of noise, dust, discharges to water and visual intrusion. All new open cast coal mines and underground mines will generally require EIA. For clay, sand and gravel workings, quarries and peat extraction sites, EIA is more likely to be required if they would cover more than 15 hectares or involve the extraction of more than 30,000 tonnes of mineral per year. |
| | Extraction of minerals by dredging in fluvial waters | Particular consideration should be given to noise, and any wider impacts on the surrounding hydrology and ecology. EIA is more likely to be required where it is expected that more than 100,000 tonnes of mineral will be extracted per year. |
| | Deep drilling | EIA is more likely to be required where the scale of the drilling operations involves development of a surface site of more than 5 hectares. Regard should be had to the likely wider impacts on surrounding hydrology and ecology. On its own, exploratory deep drilling is unlikely to require EIA. It would not be appropriate to require EIA for exploratory activity simply because it might eventually lead to some form of permanent activity. |
| | Surface industrial installations for the extraction of coal, petroleum, natural gas, ores, or bituminous shale | The main considerations are likely to be the scale of development, emissions to air, discharges to water, the risk of accident and the arrangements for transporting the fuel. EIA is more likely to be required if the development is on a major scale (site of 10 hectares or more) or where production is expected to be substantial (e.g. more than 100,000 tonnes of petroleum per year). |
| Energy Industry | Power stations | EIA will normally be required for power stations which require approval from the Scottish Ministers (i.e. those with a thermal output of more than 50 MW). EIA is unlikely to be required for smaller new conventional power stations. Small stations using novel forms of generation should be considered carefully in line with guidance in National Planning Policy Guideline 6 on Renewable Energy, and Planning Advice Note 45 on Renewable Energy Technologies. The main considerations are likely to be the level of emissions to air, arrangements for the transport of fuel and any visual impact. |
| | Surface storage of fossil fuel and natural gas, underground storage of combustible gases, storage facilities for petroleum, petrochemical and chemical products | In addition to the scale of the development, significant effects are likely to depend on discharges to water, emissions to air and risk of accidents. EIA is more likely to be required where it is proposed to store more than 100,000 tonnes of fuel. Smaller installations are likely to require EIA where hazardous chemicals are stored. |

Continued

| Class of project | Project type | Threshold/criteria |
|---|---|---|
| Energy Industry | Installations for the processing and storage of radioactive waste | EIA will normally be required for new installations whose primary purpose is to process and store radioactive waste, and which are located on sites not previously authorised for such use. In addition to the scale of any development, significant effects are likely to depend on the extent of routine discharges of radiation to the environment. In this context EIA is unlikely to be required for installations where the processing or storage of radioactive waste is incidental to the main purpose of the development (e.g. installations at hospitals or research facilities). |
| | Installations for hydroelectric energy production | In addition to the physical scale of the development, particular regard should be had to the potential wider impacts on hydrology and ecology. EIA is more likely to be required for new hydro-electric developments which have more than 5 MW of generating capacity. |
| | Wind farms | The likelihood of significant effects will generally depend on the scale of the development and its visual impact, as well as potential noise impacts. EIA is more likely to be required for commercial developments of five or more turbines, or more than 5 MW of new generating capacity. |
| Industrial and Manufacturing Development | Industrial and manufacturing development | New manufacturing or industrial plants of the types listed in the Regulations may well require EIA if the operational development covers a site of more than 10 hectares. Smaller developments are more likely to require EIA if they are expected to give rise to significant discharges of waste, emission of pollutants or operational noise. Among the factors to be taken into account in assessing the significance of such effects are:<br>• whether the development involves a process designated as a 'scheduled process' for the purpose of air pollution control;<br>• whether the process involves discharges to water which require the consent of the Scottish Environment Protection Agency;<br>• whether the installation would give rise to the presence of environmentally significant quantities of potentially hazardous or polluting substances;<br>• whether the process would give rise to radioactive or other hazardous waste;<br>• whether the development would fall under Council Directive 96/82/EC on the control of major accident hazards involving dangerous substances (COMAH).<br>However, the need for a consent under other legislation is not itself a justification for EIA. |
| Infrastructure Development | Industrial estates | EIA is more likely to be required if the site area of the new development is more than 20 hectares. In determining whether significant effects are likely, particular consideration should be given to the potential increase in traffic, emissions and noise. |
| Continued | | |

| Class of project | Project type | Threshold/criteria |
|---|---|---|
| Infrastructure Development | Urban development projects (including the construction of shopping centres and car parks, sports stadiums and multiplex cinemas) | In addition to the physical scale of such developments, particular consideration should be given to the potential increase in traffic, emissions and noise. EIA is unlikely to be required for the redevelopment of land unless the new development is on a significantly greater scale than the previous use, or the types of impact are of a markedly different nature or there is a high level of contamination. Development proposed for sites which have not previously been intensively developed is more likely to require EIA if:<br>• the site area of the scheme is more than 5 hectares; or<br>• it would provide a total of more than 10,000 m² of new commercial floorspace; or<br>• the development would have significant urbanising effects in a previously non-urbanised area (e.g. a new development of more than 1,000 dwellings). |
| | Intermodal transhipment facilities, and intermodal terminals | In addition to the physical scale of the development. particular impacts for consideration are increased traffic, noise, emissions to air and water. Developments of more than 5 hectares are more likely to require EIA. |
| | Motorway service areas | Impacts likely to be significant are traffic, noise, air quality, ecology and visual impact. EIA is more likely to be required for new motorway service areas which are proposed for previously undeveloped sites and if the proposed development would cover an area of more than 5 hectares. |
| | Construction of roads, railways (including elevated and underground) and tramways | For linear transport schemes, the likelihood of significant effects will generally depend on the estimated emissions, traffic, noise and vibration and degree of visual intrusion and impact on the surrounding ecology. EIA is more likely to be required for the construction or improvement of railways and local roads where the new development is over 2 km in length. |
| | Construction of inland waterways and canalisation | The likelihood of significant impacts is likely to depend primarily on the potential wider impacts on the surrounding hydrology and ecology. EIA is more likely to be required for the construction or improvement of over 2 km of canal. |
| | Flood relief works | The impact of flood relief work is especially dependent upon the nature of the location and the potential effects on the surrounding ecology and hydrology. Schemes for which the area of the works would exceed 5 hectares or more than 2 km long would normally require EIA. |
| | Construction of airfields | The main impacts to be considered in judging significance are noise, traffic generation and emissions. New permanent airfields will normally require EIA, as will major works (such as new runways or terminals with a site area of more than 10 hectares) at existing airports. Smaller scale development at existing airports is unlikely to require EIA unless it would lead to significant increases in air or road traffic. |

Continued

| Class of project | Project type | Threshold/criteria |
|---|---|---|
| Infrastructure Development | Construction of harbours and port installations, including fishing harbours | Primary impacts for consideration are those on hydrology, ecology, noise and increased traffic. EIA is more likely to be required if the development is on a major scale (e.g. would cover a site of more than 10 hectares). Smaller developments may also have significant effects where they include a quay or pier which would extend beyond the high water mark or would affect wider coastal processes. |
| | Dams and other installations designed to hold water or store it on a long-term basis | In considering such developments, particular regard should be had to the potential wider impacts to the hydrology and ecology, as well as to the physical scale of the development. EIA is likely to be required for any major new dam (e.g. where the construction site exceeds 20 hectares). |
| | Installation of oil pipelines, gas pipelines and long-distance aqueducts (including water and sewerage pipelines) | For underground pipelines, the major impact to be considered will generally be the disruption to the surrounding ecosystems during construction, while for overground pipelines visual impact will be a key consideration. EIA is more likely to be required for any pipeline over 5 km long. EIA is unlikely to be required for pipelines laid underneath a road, or for those installed entirely by means of tunneling. |
| | Coastal work to combat erosion and maritime works capable of altering the coast | The impact of such works will depend largely on the nature of the particular site and the likely wider impacts on natural coastal processes outside of the site. EIA will be more likely where the area of the works would exceed 1 hectare. |
| | Groundwater abstraction and artificial groundwater recharge schemes; works for the transfer of water resources between river basins | Impacts likely to be significant are on hydrology and ecology. Developments of this sort can have significant affects on environments some kilometres distant. This is particularly important for wetland and other sites where the habitat and species are particularly dependent on an aquatic environment. EIA is likely to be required for developments where the area of the works exceed 1 hectare. |
| | Ski-runs, ski-lifts and cable cars and associated developments | EIA is more likely to be required if the development is over 500 metres in length, or if it requires a site of more than 5 hectares. In addition to any visual or ecological impacts, particular regard should also be had to the potential traffic generation. |
| | Marinas | In assessing whether significant effects are likely, particular regard should be had to any wider impacts on natural coastal processes outside the site, as well as the potential noise and traffic generation. EIA is more likely to be required for large new marinas, for example where the proposal is for more than 300 berths (seawater site) or 100 berths (freshwater site). EIA is unlikely to be required where the development is located solely within an existing dock or basin. |

| Class of project | Project type | Threshold/criteria |
|---|---|---|
| Tourism and Leisure | Holiday villages and hotel complexes outside urban areas and associated developments; permanent camp sites and caravan sites; theme parks | In assessing the significance of tourism development, visual impacts, ecosystems and traffic generation will be key considerations. The effects of new theme parks are more likely to be significant if it is expected that they will generate more than 250,000 visitors per year. EIA is likely to be required for major new tourism and leisure developments which require a site of more than 10 hectares. In particular, EIA is more likely to be required for holiday villages or hotel complexes with more than 300 bed spaces, or for permanent camp sites or caravan sites with more than 200 pitches. |
| | Golf courses | New 18 hole golf courses are likely to require EIA. The main impacts are likely to be those on the surrounding hydrology, ecosystems and landscape, as well as those from traffic generation. Developments at existing golf courses are unlikely to require EIA. |
| Other Projects | Permanent racing and test tracks for motorised vehicles | Particular consideration should be given to the size, noise impacts, emissions and the potential traffic generation. EIA is more likely to be required for developments with a site area of 20 hectares or more. |
| | Installations for the disposal of non-hazardous waste | The likelihood of significant effects will generally depend on the scale of the development and the nature of the potential impact in terms of discharges, emissions or odour. For installations (including landfill sites) for the deposit, recovery and/or disposal of household, industrial and/or commercial wastes (as defined by the Controlled Waste Regulations 1992) EIA is more likely to be required where new capacity is created to hold more than 50,000 tonnes per year, or to hold waste on a site of 10 hectares or more. Sites taking smaller quantities of these wastes, sites seeking only to accept inert wastes (demolition rubble etc.) or Civic Amenity sites are unlikely to require EIA. |
| | Sludge deposition sites (sewage sludge lagoons) | Similar considerations will apply for sewage sludge lagoons as for waste disposal installations. EIA is more likely to be required where the site is intended to hold more than 5000 $m^3$ of sewage sludge. |
| | Storage of scrap iron, including scrap vehicles | Major impacts are likely to be discharges to soil, site noise and traffic generation. EIA is more likely to be required where it is proposed to store scrap on an area of 10 hectares or more. |
| | Waste-water treatment plants | Particular consideration should be given to the size, treatment process, pollution and nuisance potential, topography, proximity of dwellings, and the potential impact of traffic movements. EIA is more likely to be required if the development would be on a substantial scale (e.g. site area of more than 10 hectares) or if it would lead to significant discharges (e.g. capacity exceeding 100,000 population equivalent). EIA should not be required simply because a plant is on a scale which requires compliance with the Urban Waste Water Treatment Directive 91/271/EEC. |

## Selection Criteria for Screening Schedule 2 Development
(Annexe B Circular 15/1999)

### 1. Characteristics of development

The characteristics of development must be considered having regard, in particular, to:

a. the size of the development;
b. the cumulation with other development;
c. the use of natural resources;
d. the production of waste;
e. pollution and nuisances;
f. the risk of accidents, having regard in particular to substances or technologies used.

### 2. Location of development

The environmental sensitivity of geographical areas likely to be affected by development must be considered, having regard, in particular, to:

a. the existing land use;
b. the relative abundance, quality and regenerative capacity of natural resources in the area;
c. the absorption capacity of the natural environment, paying particular attention to the following areas:
  i) wetlands;
  ii) coastal zones;
  iii) mountain and forest areas;
  iv) nature reserves and parks;
  v) areas classified or protected under Member States' legislation; areas designated by Member States pursuant to Council Directive 79/409/EEC on the conservation of wild birds and Council Directive 92/43/EEC on the conservation of natural habitats and of wild fauna and flora;
  vi) areas in which the environmental quality standards laid down in Community legislation have already been exceeded;
  vii) densely populated areas;
  viii)landscapes of historical, cultural or archaeological significance.

### 3. Characteristics of the potential impact

The potential significant effects of development must be considered in relation to criteria set out under paragraphs 1 and 2 above, and having regard, in particular, to:

a. the extent of the impact (geographical area and size of the affected population);
b. the transfrontier nature of the impact;
c. the magnitude and complexity of the impact;
d. the probability of the impact;
e. the duration, frequency and reversibility of the impact.

# Annexe 5   References and Annotated Bibliography

## Annotated Bibliography of References in Text

(1) SNH (1996) *The Design and Build Process for Trunk Roads: A Guide for SNH*, prepared for SNH by ERM
  Internal guidance note for SNH staff on the design and build process for trunk roads, utilising SO Industry Dept. National Roads Directorate techniques and explaining the role of environmental assessment in the process; see section 6 of the Guidance Note.

(2) The Scottish Office (Sept 1996) *The Assessment of Trunk Road Projects: Consultation Process Between the National Roads Directorate and Scottish Natural Heritage: Consultation Guidance Note* (Consultation Draft)
  Internal guidance note for SNH staff on established consultation arrangements between the National Roads Directorate and SNH, including agreed procedures for establishing whether environmental assessment is required and the procedures for consultation where assessment is relevant.

(3) Department of the Environment (1989) *Environmental Assessment: A Guide to the Procedures*, HMSO, London
  One of the first authoritative, and still widely used, guidance publications on the environmental assessment process. Tends to focus on procedures and statutory requirements rather than non-statutory procedures which are good practice.

(4) Commission of the European Communities (1985) *Council Directive on the Assessment of the Effects of Certain Public and Private Projects on the Environment* (85/337/EEC)
  The EC Directive which triggered statutory environmental assessment procedures for the first time in the UK. It is still the basis of all EC and UK legislation but is currently being updated and extended in the light of practice.

(5) Environmental Assessment (Scotland) Regulations 1988 (SI 1988 No. 1221)
  The Regulations which introduced many of the requirements for environmental assessment in Scotland in respect of a wide range of projects, in 1988, to comply with the EC Directive. Other Regulations were introduced at the same time and subsequently; see Annexe 2 of this Handbook.

(6) Town and Country Planning (Scotland) Act 1997
  The basic legislative provision for town and country planning in Scotland. The Act requires all new development to obtain planning permission and the process of obtaining permission is used to apply the environmental assessment regulations through the EASR 88, the Amendment Regulations of 1994 and by the Town and Country Planning (General Development Procedure) (Scotland) Order 1992.

Town and Country Planning (General Development Procedure) (Scotland) Order, 1992.

This General Development Order contains provisions for requiring further information on planning applications under Articles 6 and 13 and for the Scottish Ministers to issue Directions about Environmental Assessment under Articles 16 and 19.

(7) Electricity (Supply) Acts 1882–1936

Statutory provision requiring the consent of the Scottish Ministers for certain types of electricity works; the EASR 88 apply to these consents.

(8) Schedule to the Electric Lighting Clauses Act 1899

Statutory provision requiring the consent of the Scottish Ministers for certain types of electricity works; the EASR 88 apply to these consents.

(9) Electricity (Scotland) Act 1979

Frequently used statutory provision requiring the consent of the Scottish Ministers for certain types of electricity works; the EASR 88 apply to these consents.

(10) New Towns (Scotland) Act 1968

An Act making statutory provision for New Town Corporations in Scotland to obtain the consent of the Scottish Ministers, and in some cases the planning authority, for new town developments. Now superseded; the EASR 88 applied to these consents.

(11) Land Drainage (Scotland) Act 1968

Statutory provision requiring the consent of the Scottish Ministers for improvement orders for drainage works on agricultural land in Scotland, the EASR 88 apply to these orders, see sections C.32 and C.33 of the Local Authorities Handbook.

(12) Roads (Scotland) Act 1984

Statutory provision requiring the consent of the Scottish Ministers or enabling the Scottish Ministers directly to carry out works for new roads or bridges, and road or bridge improvements; the EASR 88 apply to these consents.

(13) New Roads and Street Works Act 1991

Section 42 requires environmental assessment in every case for a 'special road' including motorways and both public and (as amended by regulation 4 of the Amendment Regulations SI 1994 No. 1221) privately financed toll roads.

(14) Scottish Executive Development Department PAN 58 (1999) *Environmental Impact Assessment*

This Planning Advice Note provides background information and advice on good practice in the EIA process to supplement the legal, administrative and policy advice in Circular 15/1999.

(15) Transport and Works Act 1992

Section 14 of the Act ensures that all Schedule 1 and Schedule 2 projects likely to have significant effects on the environment which are to be authorised or consented under the Act are subject to the Environmental Assessment procedures. These projects may include a wide range of large scale or linear infrastructural works.

(16)   Parliamentary Standing Order No. 37A
These Parliamentary procedures ensure that all relevant Schedule 1 and Schedule 2 projects which are to be authorised by Parliament directly are subject to the Environmental Assessment procedures, usually at Committee stage. These projects may include a wide range of usually large scale infrastructural works.

(17)   Scottish Executive Development Department Circular 15/1999
*The Environmental Impact Assessment (Scotland) Regulations 1999*
This Circular provides comprehensive guidance on the EIA process with particular emphasis on projects requiring planning permission and those requiring approval under the Roads (Scotland) Act or the Land Drainage Acts.

(18)   Environmental Impact Assessment (Scotland) Regulations 1999.
See Annexe 1 above; this covers EIA procedures for projects requiring planning permission or approval under either the Roads (Scotland) Act or the Land Drainage Acts.

(19)   European Community EC Directive 97/11/EC of 3.3.97
Amending the 1985 Directive on Environmental Assessment. The key provisions of this Amending Directive are described in Section B.2 and Annexe 4 of this Handbook.

(20)   SNH PL 94/6, *SNH Guidance on the Strategic Environmental Assessment of EC Structural Funds in Scotland* Annexe 1
Internal guidance note on the Strategic Environmental Assessment of EC Structural Funds in Scotland. Funded plans are not necessarily subject to assessment at the project level so this procedure provides an overview of likely significant effects of the application of structural funds in Scotland on the natural heritage.

(21)   Institute of Environmental Assessment (1995) *Guidelines for Baseline Ecological Assessment*, Spons
The most authoritative and widely recognised and adopted best practice guide for ecological baseline studies in the environmental assessment process. Recommended as good practice guidance by all relevant Institutes. See particularly Appendix 2 of this Handbook.

(22) a)   Institute of Environmental Assessment/The Landscape Institute (2002)
*Guidelines for Landscape and Visual Impact Assessment*, Second Edition, Spons
The most authoritative and widely recognised and adopted best practice guide for landscape and visual impact assessment in the environmental assessment process. Unlike (21) above, these guidelines cover the whole of the environmental assessment process. Recommended as good practice guidance by all relevant Institutes, supported and partly funded by SNH. See particularly Appendix 1 of this Handbook.

b) Countryside Agency and SNH (2002) *Landscape Character Assessment, Guidance for England and Scotland*, prepared by Carys Swanwick, Department of Landscape, University of Sheffield and Land Use Consultants
This document consolidates the latest thinking on the principles and practice of landscape character assessment, and is a useful basis from which to understand and use the Scotland-wide suite of Landscape Character Assessments. Each SNH office should have the Landscape Character Assessments for the relevant local authority area or areas.

c) SNH, *Marine Aquaculture and the Landscape: The siting and design of marine aquaculture developments in the landscape*

d) SNH, *Guidelines on the Environmental Impacts of Windfarms and Small Scale Hydroelectric Schemes*

e) SNH, *Minerals and the Natural Heritage in Scotland's Midland Valley*

(23)   Department of the Environment (1995) *Preparation of Environmental Statements for Planning Projects That Require Environmental Assessment: A Good Practice Guide*, HMSO

> Guide to projects requiring Environmental Assessment under the UK Town and Country Planning legislation. The first part of the book covers scoping and the definition of requirements for baseline studies, prediction of impacts and formulation of mitigation measures. The second part describes the preparation of an Environmental Statement. Excellent manual, particularly valuable for local authorities, developers and consultants.

(24)   Conservation (Natural Habitats etc.) Regulations 1994

> The statutory provisions introduced in the UK to give effect to the international obligations of the Habitats and Birds Directives, see particularly para C.15 and section E.2 of this Handbook.

(25)   RSPB (1995) *Wildlife Impact: The Treatment of Nature Conservation in Environmental Assessment*, RSPB, Sandy, Beds

> An independent analysis of some 38 environmental statements for a wide variety of project types in Great Britain, from 1988 to 1995, published by the RSPB. Generally the research conclusions were consistent with others showing an improvement since 1992 but still many weaknesses in the way that the process is carried out. The research showed how consultation responses, e.g. from SNH, could strongly influence decisions, especially where the Statement was poor.

(26)   Institute of Environmental Assessment (1993) *Practical Experience of Environmental Assessment in the UK*, IEA

> An authoritative, thorough but relatively succinct overview of practice 1988 to 1993, by the IEA itself. Many shortcomings in practice are highlighted emphasising how reliant the statutory procedures are on integrity and quality of the environmental statements.

(27)   a) Jones CE, Lee N and Wood C (1991) *UK Environmental Statements 1988–1990: An Analysis*. Occasional Paper 29, University of Manchester EIA Centre

   b) Lee N and Colley R (1990) *Reviewing the Quality of Environmental Statements*. Occasional Paper 24, Department of Planning and Landscape, University of Manchester

   c) Lee N and Colley R (1992) *Reviewing the Quality of Environmental Statements*. Occasional Paper 24, Department of Planning and Landscape, University of Manchester, Second Edition

   d) Environmental Impact Assessment Centre, University of Manchester, for Dept of Environment (1991) *Monitoring Environmental Assessment and Planning*, HMSO

> A series of readable, well researched studies analysing the effectiveness of

environmental assessment for a variety of clients and for University research purposes. The Centre is highly regarded internationally as one of excellence in the field of environmental assessment.

(28)   Environmental Information Regulations
   The UK regulations introduced to give effect to the EC Directive 90/313/EC on Freedom of Access to Information on the Environment.

(29)   EC Directive 90/313/EC on Freedom of Access to Information on the Environment
   EC Directive ensuring that the public has a right of access to environmental information, applied in the UK through the Regulations at 28 above.

(30)   *SNH Staff Guidance on Freedom of Information on the Environment*, July 2004
   Detailed internal guidance on freedom of access to environmental information and the implications of the EC Directive and UK Regulations at 28 and 29 above.

(31)   Natural Heritage (Scotland) Act 1991
   SNH's founding legislation, requiring SNH to give advice, on request, to planning authorities and others about the natural heritage.

(32)   Wood C and Jones C (1995) *The Effect of Environmental Assessment on Planning Decisions*. Workshop at Manchester University, July
   A useful and reasonably succinct resume of the relationship between the environmental assessment process and decision making in planning authorities. Scotland is not well covered.

(33)   *This Common Inheritance* (1990) HMSO
   The foundation on which the current UK government policy approaches to sustainable development and environmental planning principles, including environmental assessment and strategic environmental appraisal have been built.

(34)   SNH Discussion Paper (1996) *The Precautionary Principle–Step by Step Guide* Policy Guidance Note 96/1
   SNH policy paper explaining an approach to the application of the Precautionary Principle.  See also SNH (2000) 'Applying the Precautionary Principle to decisions on the natural heritage'

(35)   Scottish Executive  Circular June 2000 replacing Circular 6/1995 *The Conservation (Natural Habitats Etc.) Regulations 1994*
   This Circular provides procedural and policy guidance on the Habitats Regulations 1994, and specifically indicates that any project likely to have a significant effect on a Natura 2000 (European) Site, whether fully designated or not, should normally be subject to the Environmental Assessment process. The Circular also explains how this differs from the appropriate assessment undertaken by the Competent Authority under the Habitats Regulations.

(36)   Catlow J and Thirlwall G (1997) *Environmental Assessment*. Report to Department of the Environment, HMSO
   The first officially commissioned report on environmental assessment in the UK, now difficult to obtain but still remarkably relevant to present day issues. It

would have formed the basis of UK legislation had this not been overtaken by the EC Directive coming into force in July 1988.

(37)   Food and Environment Protection Act 1985
Amongst many other things this Act provides the licensing system for any deposits or structures on the sea bed, below mean high water mark.

(38)   *Likely Significant Effect* (1999)
A paper produced jointly by the country agencies for nature conservation, available from SNH.


## Further Annotated Bibliography

Department of Trade and Industry (1992) *Guidelines for the Environmental Assessment of Cross-country Pipelines*, Department of Trade and Industry
Department of Trade and Industry authoritative guidance on environmental assessment of pipeline projects, including when the process will be applicable.

Department of the Environment (1995) *The Environmental Impact of Leisure Activities Fourth Report*, Volume 1, HMSO
This report looks at the impact of the visiting public and their leisure pursuits on the UK rural environment.

Department of Transport (1993) *Manual of Environmental Appraisal*, Volume 11, *Design Manual for Roads and Bridges*, Department of Transport
Developed from the Department of Transport's 'Manual of Environmental Assessment', which was the first detailed method prescription for environmental assessment for a particular project type in the UK. This is a comprehensive, thorough, detailed and indispensable guide to impact assessment for roads and related works. However, many of its standards and principles are equally applicable to other types of project, especially other linear projects. This is a key reference for engineers and consultants. It is very widely used in practice and must be followed meticulously by engineers and consultants working on all Scottish Office Road Schemes.

Environment Agency (1996) *Environmental Assessment: Scoping Handbook for Projects*, Environment Agency
Invaluable aid for all practitioners involved in assessments where the Agency (or SEPA) are a key consultee but, despite the generality of the title, it is exclusively water-related.

Forestry Commission, *Environmental Impact Assessment of Forestry Projects: Undertaking an Environmental Impact Assessment in forestry and preparation of an Environmental Statement.* At
www.forestry.gov.uk/website/pdf.nsf/pdf//wgseia.pdf/$file/wgseia.pdf

Glasson J, Therivel R and Chadwick A (1994) *Introduction to Environmental Impact Assessment*, UCL Press.
A valuable and readable introductory textbook, by authoritative authors.

Harwood, R (2004) Environmental Impact Assessment: What's Next? *Journal of Planning and Environmental Law*, September, 1161–1175

> Useful quasi-legal update of the EIA process and where it may be going; good references and analysis of EIA court cases overseas including outwith the EC.

Institute of Environmental Assessment, *Digest of Environmental Statements 1996, UK and Europe*, Sweet & Maxwell

> Unique compendium of UK environmental statements, drawing on the resources of the IEA's comprehensive library of statements. Each statement entry outlines a brief planning history or summary of the assessment itself, and provides details of the authors and contributors involved. Details of those statements for which a decision is pending are included, as well as the text of the regulations under which an application has been made. The Digest is an expensive annual publication, issued as two releases per annum. Additional documentation can be purchased – full details from the IEA.

Institute of Environmental Assessment (1993) *Guidelines for the Environmental Assessment of Road Traffic*, Institute of Environmental Assessment

> The third of the Institute of Environmental Assessment's guidance publications; see also references 21 and 22 above. The most authoritative and widely recognised and adopted best practice guide for traffic impact assessment in the environmental assessment process, where the project itself is not a road or traffic scheme. The guidelines cover the whole of the environmental assessment process. Recommended as good practice guidance by all relevant institutes.

Morgan RK (1997) *Environmental Impact Assessment: A Methodological Approach*, Chapman & Hall

> This text provides a relatively straightforward introduction to methodologies and approaches.

Morris P and Therivel R (eds) (1995) *Methods of Environmental Impact Assessment*, UCL Press

> A practical, up-to-date explanation of and guide to how statements are, and should be, carried out for specific environmental components (e.g. air, water, ecological systems, socio-economic systems). For each component, it includes a discussion of relevant regulations and standards, how baseline surveys are conducted, how impact predictions are made, what mitigation measures can be used, how the effectiveness of such measures should be monitored, and the limitations of the methods.

Roe D, Dalal-Clayton B and Hughes R (1995) *A Directory of Impact Assessment Guidelines*, IIED

> This directory includes guidelines for environmental, health and social impact assessment, drawing together documents from national governments, development banks, donor agencies, international organisations and NGOs. Over 450 documents are cited, with 150 abstracts, covering key sectors in every region of the world.

Scottish Natural Heritage (in press) *An Introduction to Woodlands and Forestry*, SNH

> Provides important background material for use in the assessment of environmental statements, especially in respect of the Forestry EIA procedures, but would be relevant to any project that affected woodlands.

Smith LG (1996) *Impact Assessment and Sustainable Resource Management*, Addison Wesley Longman
> This book explores some of the fundamental issues associated with impact assessment, identifies current strengths and weaknesses, and suggests changes necessary to ensure impact assessment contributes fully to the achievement of sustainable resource management.

Therivel R (1992) *Strategic Environmental Assessment*, Earthscan
> Strategic Environmental Assessment is described as being the developing method of EIA aimed at ensuring that projects involving strategic decisions are based on a full understanding of their likely environmental consequences. Using UK based but globally applicable examples, this book reviews Strategic Environmental Assessment in relation to other tactics for environmental protection.

Therivel R and Rosario Paridario M (1996) *The Practice of Strategic Environmental Assessment*, Earthscan
> Provides a unique analysis of Strategic Environmental Assessments which have been undertaken, drawing on a variety of methods and circumstances to illustrate how best practice can be achieved, and providing inspiration for those considering studying, commissioning or carrying out an Strategic Environmental Assessment. This is probably, academically,  the most influential book publication on Strategic Environmental Assessment so far.

Treweek J (1999) *Ecological Impact Assessment*, Blackwell Science
> Recent publication addressing ecological impact assessment principles and practice.

Tromans S and Fuller K (2003) *Environmental Impact Assessment, Law and Practice*, Butterworths Environmental Law Series, Lexis Nexis Butterworths in association with the Institute of Environmental Management and Assessment
> Authoritative book on law and practice, comprehensive of issues, but many Scottish statutes are missing and the book is unreliable for use in Scotland if relying on legislation.

Wathern P (ed.) (1997) *Environmental Impact Assessment Legislation in the EC*, Wiley
> The first published analysis of the various methods by which the EC Directive on EIA has been implemented in each of the member states. Differences between procedures and practice within and between countries are highlighted, providing a valuable context for assessments in the UK and a useful source of information.

Wathern P (ed.) (1990) *Environmental Impact Assessment Theory and Practice*, Chapman & Hall
> A review which covers technical aspects and the effects of environmental assessment on the decision making process. A major textbook which is soundly based on theory and practice from an authoritative author.

Weisner D (1995) *Environmental Impact Assessment: The Environmental Impact Assessment Process, What Is It and How to Do One*, Prism Press.
> Useful popular guide which seeks to provide non-professional groups and individuals with the tools to make worthwhile and substantive comment on a development proposal.

Wood C (1996) *Environmental Impact Assessment, A Comparative Review*, Addison Wesley Longman

> An authoritative, international review by the well respected EIA Centre, Department of Planning and Landscape, University of Manchester. The book compares systems used in the UK, US, Netherlands, Canada, Australia and New Zealand. Standard procedures are described; each step of the process is discussed; best current practice is explored and the future direction is surveyed. A significant work internationally.

## Other Potentially Useful References

Beanlands GE and Duinker PN (1983) *An Ecological Framework for Environmental Impact Assessment in Canada*. Institute for Resource and Environmental Studies, Dalhousie University

Box JD and Forbes J (undated) *Ecological Considerations in the Environmental Assessment of Road Proposals*, unpublished draft document. English Nature

Countryside Commission (1991) *Environmental Assessment: The treatment of landscape and countryside issues*, Countryside Commission Publications

Department of the Environment (1991) Consultation Paper: *Environmental Assessment and Private Bill Procedures*, PDC 4, DoE, London

Scottish Office Circular 26/1991 *Environmental Assessment and Private Legislation Procedures*

Scottish Office Circular 3/1991 *Electricity Generating Stations and Overhead Lines: Permitted Development for Electricity Undertakings*

Scottish Office (1994) National Planning Policy Guidelines (NPPG) No. 1. *The Planning System*

Scottish Office Circular 26/1988 *Environmental Assessment of Projects in Simplified Planning Zones and Enterprise Zones*

Spellerberg IF (1992) An investigation into the nature and use of ecology in EIAs. *British Ecological Society Bulletin*, 23, 38–45.

Wathem P (ed.) (1988) *Environmental Impact Assessment: Theory and Practice*, Unwin Hyman, London

# Annexe 6    Historical Development of Environmental Assessment in Scotland

## First UK Examples in Scotland

**An.6.1**   The first examples of Environmental Assessment in the UK occurred in Scotland, in the early 1970s, in relation to the major infrastructure developments for North Sea oil and gas installations on the Firth of Forth. These commendable early attempts to use the process of Environmental Assessment were entirely voluntary. Environmental Assessment was not introduced as a statutory requirement until 1988. This section briefly outlines the historical development of Environmental Assessment, internationally and nationally, to provide an understanding of why the process was introduced and its original intentions and to shed light on the current approaches to Environmental Assessment.

## International Recognition of the Need for Environmental Assessment

**An.6.2**   A number of factors contributed to the international recognition of the need for and the development of Environmental Assessment. These included:

- the apparent failure of traditional project appraisal techniques such as Cost/Benefit Analysis (CBA) to account for intangible environmental effects;

- the growth of environmental awareness particularly in the United States;

- the recognition that the efficiency and profitability of some commercial projects had been affected by the consequent environmental changes they brought about and that unforeseen risks associated with such impacts could be environmentally damaging and commercially unacceptable;

- a number of widely reported disasters which highlighted the risks to the environment from human activities, such as: the mercury poisoning from a factory in Minamata, Japan (1952–1960); recognition of the effects of the Aswan Dam on the fertility of the Nile valley; and the Torrey Canyon oil spill in the English Channel (1967).

## US Legislation 1969

**An.6.3**   The first legislation requiring environmental assessment was enacted in the US in 1969. The National Environmental Policy Act was adopted by the Nixon administration in 1970. Amongst other things, the Act required federal agencies to include in every recommendation for legislation, and other major federal actions that may significantly affect the quality of the human environment, a detailed statement to assess:

- the environmental impacts of the proposed action;
- any unavoidable adverse environmental effects should the proposal go ahead;
- alternatives to the proposed action;
- the relationship between local short-term uses of the environment and the maintenance and enhancement of long-term productivity;

- any irreversible and irretrievable commitments of resources which would be involved.

**An.6.4**  Despite considerable teething problems many of the NEPA's ideas and provisions became widely accepted and it formed a recognised model for Environmental Assessment adopted or adapted by a number of countries around the world.

## Early UK Initiatives

**An.6.5**  In the UK the DoE commissioned a report in 1974 which was intended to examine the scope for and feasibility of introducing Environmental Assessment into UK procedures. The report was produced by John Catlow and Geoffrey Thirlwall in 1976, and eventually published by DoE in 1977 (36). The recommendations of that report were progressed so slowly that they were eventually overtaken by the EC Directive requiring Member States to introduce domestic legislation to comply. This, effectively, led to the implementation of many of the recommendations in the 1977 report, but not all of the report's main conclusions have been adopted in statutory form, although many remain relevant as good practice rather than mandatory requirements.

**An.6.6**  For example, the 1977 report recognised that analysis should commence early in the preparation of the development proposal to be useful as a design tool and to examine alternatives; that analysis should include economic and social impacts as well as those affecting the physical environment; that the study should be carried out by a team of experts, from a wide range of disciplines, and should be supervised by the planning authority and developer in cooperation; and that a responsible authority should determine what environmental impacts are likely to be relevant and therefore should be included in the analysis.

**An.6.7**  The 1977 report envisaged only a small number of projects ever being appropriate for Environmental Assessment but it soon became evident that many more projects would have to comply.

## The First EC Directive

**An.6.8**  The EC Directive itself had proved to be controversial. It had been circulated as a draft as early as 1980 but there had been severe delays in reaching a standard and policy acceptable to all Member States, some of whom already had Environmental Assessment provisions of their own. Eventually, compromises were found and Environmental Assessment procedures were formally introduced into the European Community through the Directive 85/337/EEC 'The Assessment of the Effects of Certain Public and Private Projects on the Environment'. It allowed 3 years for Member States to implement the proposals through national legislation. UK Regulations were first introduced just after the compliance date, in July 1988, but gaps in compliance have led to a continuing series of further Regulations, those relevant in Scotland being listed at the front of this Handbook.

**An.6.9**  The principal aims of the Directive were:

- to ensure that the environmental consequences of new development were known and taken into account before any consent could be granted; and

- to encourage developers to consider environmental concerns from the earliest stage of project planning and design, when potentially adverse effects can be most effectively and economically addressed.

**An.6.10** It follows from this second objective that developers were responsible for having the analysis carried out, and needed to promote interaction between project design and environmental concerns.

**An.6.11** The Directive consisted of 14 articles and three annexes. The major provisions are listed below:

- Member States must adopt *'all measures necessary'* to ensure that *'before consent is given, projects likely to have significant effects on the environment by virtue, among other things, of their nature, size or location are made subject to an assessment with regard to their environmental effects'*.

- Requirements may be integrated into the existing consent procedures of individual states which were allowed considerable discretion in implementation.

- Exemptions from Environmental Assessment requirements could be made in exceptional circumstances. (In the UK a number of Ministry of Defence projects have been exempted on the grounds of national security.)

- The types of development affected were those which were *'likely to have significant effects on the environment'* and were listed in 2 Annexes to the Directive:
  - Schedule 1 projects, which should always be subject to Environmental Assessment and
  - Schedule 2 projects, which may be subject to Environmental Assessment *'if their characteristics so require'*.

- Member States were required to develop criteria for deciding when projects listed in Schedule 2 should be subject to Environmental Assessment and to review these criteria periodically.

- The information which should be included in an Environmental Statement was specified, in Annexe III, but the Directive did not prescribe assessment methods.

## Experience of Statutory Environmental Assessment in the UK and Scotland

**An.6.12** Over 1000 environmental statements were submitted in the first 5 years following the introduction of statutory Environmental Assessment in the UK, in 1988. By February 1999, 347 Environmental Statements had been submitted in Scotland, in respect of all kinds of projects that are subject to Environmental Assessment. 37 of these related to Schedule 1 projects; the others related mainly to minerals (92), waste (66), wind energy (27) and urban projects (32). Self-evidently, this far exceeds the number envisaged by the Government and means that Environmental Assessment is now a well-established and by no means uncommon procedure. The number of Environmental Assessment cases is likely to increase owing to the new Regulations in 1999 widening the scope of projects requiring EIA.

## Standards and Effectiveness of Environmental Assessment

**An.6.13**  The debate about Environmental Assessment used to be focused on the number of Environmental Assessment cases that ought to be subject to assessment and whether there is a need for Environmental Assessment. However, it has extended to include a debate about the standards of Environmental Statements and the effectiveness of the procedures. Important research projects, separately undertaken on behalf of the IEA (26), DoE (27) and RSPB (25), have exhibited remarkably consistent conclusions which include a distinct improvement in the quality of Environmental Statements since about 1992.

**An.6.14**  This appears to be the direct consequence of several important factors, namely:

a. the wider availability and use of published good practice guidance;

b. the increasing level of experience of Environmental Assessment particularly in consultancies that have prepared several Environmental Statements;

c. a wider recognition that Environmental Assessment can be a useful and positive contribution to project design and management;

d. the increasing proportion of Environmental Statements that have been subject to prior scoping and consultation;

e. the increasing experience of developers, Competent Authorities and consultees in dealing with Environmental Assessment and knowing what information to require and how to deal with it.

**An.6.15**  The best Environmental Statements have been those which involved:

- thorough scoping and continuing consultation;
- experienced assessors working in well co-ordinated multi-disciplinary teams with qualified experts dealing with specific topics;
- thorough survey and diligent research to provide comprehensive and up-to-date information based on standard survey methods;
- objective and impartial analysis of information using good practice techniques;
- clear identification of the nature, scale and significance of all relevant impacts;
- acknowledging limitations in data and understanding of impacts;
- a clear description of all mitigating measures, their effects and effectiveness and how they would be guaranteed;
- a commitment to mitigation, monitoring review and remedial procedures.

**An.6.16**  The poorer Environmental Statements were those that:

- failed or inadequately attempted to carry out early liaison and scoping of the issues;
- failed or inadequately attempted to maintain consultation during the whole process;
- failed to address the full scope of effects or to describe the development adequately;

- relied only on existing, often out of date information;
- failed to provide clear baseline data;
- failed to identify all relevant impacts and/or failed to indicate their nature, scale or significance;
- did not address the policy context in which the project would be determined;
- failed to identify/describe all mitigating measures and their effects and effectiveness;
- did not indicate how mitigation could be guaranteed and ignored monitoring.

**An6.17** The publication of Circular 15/1999 and PAN 58, in 1999, which contain much more comprehensive guidance on good practice, is likely to further raise the standards of assessment.

# Annexe 7    List of Principal Legal Cases Referred to

## In date order

*European Court of Justice, Aannemersbedrijf PK Kraaijeveld BV v Gedeputeerde Staten van Zuid-Holland October 24 (1996) (Dutch Dykes) Case C-72/95.*

*WWF UK Ltd and RSPB v SNH, the Secretary of State for Scotland, the Highland Council, Highlands and Islands Enterprise and the Cairngorm Chairlift Co Ltd. (Court of Session 28 October 1998)*

*Regina v St Edmundsbury Borough Council, ex parte Walton (1999) [1999 JPL 805]*

*Regina v Rochdale MBC ex parte (1) Andrew Tew, (2) George Daniel Milne, (3) Steven Garner Queens Bench Division, Sullivan J., (1999) [2000 JPL 54]*

*Berkeley v Secretary of State Environment Transport and the Regions (2000) [JPL 2001 58],*

*Regina v Rochdale MBC ex parte Milne (2000) [2001 JPL 470]*

*Regina v Secretary of State Environment Transport and the Regions ex parte Diane Barker (2001)*

*Regina on the application of Lebus v South Cambridgeshire DC (2002) [2003 JPL 466]*

*Fernback and Others v Harrow LBC (2000) [2001 EWHC Admin 278; 2002 Env LR 10]*

*Gillespie v First Secretary of State and Bellway Urban Renewal (TLR 7/4/2003) [14 LS Gaz R 30]*

*Goodman and another v Lewisham London Borough Council (2003) [TLR 21/2/03]*

# Technical Appendices

**Attachment A**
Guide to the Scoping and Review of an
Environmental Statement

**Appendix 1**
Landscape and Visual Impact Assessment

**Appendix 2**
Ecological Impact Assessment

**Appendix 3**
Earth Heritage Impact Assessment

**Appendix 4**
Assessment of Impacts on Soils

**Appendix 5**
Outdoor Access Impact Assessment

*Revised*

**Appendix 6**
Effects on the Marine Environment

# Attachment A  Guide to the Scoping and Review of an Environmental Statement

## Note on Use of Part 1 of the Guide

This is intended to assist Competent Authorities and consultees in their responses to a scoping request from a developer. It is not intended, and should not be used, as a framework to enable developers or their agents to produce a scoping report. This is a separate exercise not covered here.

For obvious reasons, the scope of topics is limited here to natural heritage issues but users are encouraged to extend/replace/adapt these issues to cover those which are relevant to them, e.g. the cultural heritage, air quality etc.

Tick appropriate boxes or circle appropriate answers and compile a letter to, or action list for a meeting with, the developer and/or Competent Authority.

### 1. Do you know the site?

**Yes** Go to question 2.

**No** Visit site as soon as possible or talk to someone who knows the site well, then, or in the meantime, go to question 2 and on the evidence available:

### 2. Could the proposal affect a natural heritage designation, including:

☐ National Park   ☐ Regional Park   ☐ Country Park   ☐ Picnic Site

☐ NSA   ☐ An Historic Garden or Designed Landscape

☐ AGLV   ☐ Other landscape designation

☐ (c)SAC   ☐ (p)SPA   ☐ Ramsar Site   ☐ SSSI

☐ MNR   ☐ LNR   ☐ Non-statutory wildlife site

**Yes** If any boxes ticked ensure developer/Competent Authority is fully aware of designation and its boundaries, interest features/value, reason and purpose of designation, conservation or management objectives etc. Go to question 3.

**No** Go to question 3.

## 3. Could the proposal affect any statutory or other important outdoor access facility including:

❏ Long distance route    ❏ Public right of way    ❏ Access area/route

Go to question 4.

## 4. Could the proposal affect species or habitats, for example, any:

❏ Protected species    ❏ (L)BAP species or habitat

Go to question 5.

## 5. Could the proposal have a significant effect on:

❏ The character, integrity or distinctiveness of the landscape?

❏ The amenity or enjoyment of the landscape experience including its wildland character?

❏ Important, typical, distinctive or otherwise important landscape features?

❏ The historical/cultural interest of the landscape?

**Yes**    Ensure the developer/Competent Authority is aware of the Landscape Character Assessment, Appendix 1 of this Handbook and other good practice guides and how they may inform the EIA process.
Go to question 6.

**No**    Go to question 6.

## 6. Could the proposal have a significant effect on:

❏ Any other natural heritage resources or

❏ Access to the countryside?

**Yes**    Ensure the developer/Competent Authority are aware of the interest , Appendix 5 of this Handbook and good practice guides that may inform the EIA process. Go to question 7.

**No**    Go to question 7.

**7. What in your view are the key environmental issues raised by the proposal? Use the table below to circle and note the important issues.**

| Receptor (What may be affected) | Issue (What the effect might be) | Will it be covered in ES? |
|---|---|---|
| People | | |
| Landscape | | |
| Visual amenity | | |
| Recreation/Access | | |
| Geology, rocks and minerals | | |
| Geomorphology, natural systems and processes | | |
| Soil | | |
| Water | | |
| Hydrology | | |
| River systems | | |
| Habitats | | |
| Plant species Animal species | | |
| Designed landscape | | |
| Cultural heritage | | |
| Built environment | | |
| Air quality/Climate | | |
| Other (specify) | | |

**8. Is there evidence that any of these issues will not be addressed (or will not be appropriately addressed) in the Environmental Statement?**

**Yes** Write to developer/Competent Authority expressing your views, copy letter to Competent Authority. Go to question 9.

**No** Go to question 9.

### 9. Do you know which person will be co-ordinating the preparation of the Environmental Statement?

**Yes** Go to question 10.

**No** Contact developer and find out. Go to question 10.

### 10. Do you know which persons will be responsible for assessing effects on specific issues of interest to you?

**Yes** Go to question 11.

**No** Contact developer and find out. Go to question 11.

### 11. Do you know of and agree with methodologies and timetables proposed for survey and assessment?

**Yes** Go to question 12.

**No** Write to developer/Competent Authority expressing your views, copy letter to Competent Authority. Go to question 12.

### 12. Will the Environmental Statement consider alternative solutions, e.g. other sites, designs or processes?

**Yes** Go to question 13.

**No** Where relevant write to developer/Competent Authority advocating consideration of alternative solutions. Go to question 13.

### 13. Will the EIA process involve consultation with other appropriate conservation/environmental bodies (e.g. RSPB, SWT)?

**Yes** Go to question 14.

**No** Write to developer/Competent Authority advocating consultation with relevant conservation bodies. Go to question 14.

### 14. Is the EA co-ordinator aware of relevant information held by you?

**Yes** Go to action point below.

**No** In response to scoping exercise, inform proponent of information held by you and the arrangements for obtaining it. Go to action point below.

**Action Point** Go back to beginning and collate all relevant points of concern and action points and communicate with developer and/or Competent Authority. Part 1 above, relating to scoping must be filled in before completing this Section.

# Part 2     Review of an Environmental Statement

**Part 1 above, relating to scoping must be filled in before completing this Section.**

**The questions below are intended to guide a consultee's or a Competent Authority's review of an Environmental Statement (ES) and the application proposal that it relates to.**

**Responses need to make clear whether they relate to the adequacy of an Environmental Statement or to the suitability of a proposal or both.**

**Use the technical guidance in Parts C and D and Appendices 1 to 6 of the Handbook to help you to decide the answers to the questions.**

*Circle the appropriate answers. Compile an action list and letter to the Competent Authority or annexe to send with your consultation response to the Competent Authority.*

- Is the purpose and rationale of the project clearly descibed along with how it would be carried out at each phase of the development?

Yes ↓    No →     Try and clarify with developer/Competent Authority, note deficiencies in ES in response to Competent Authority.

- Is the description of the receiving environment accurate?

Yes ↓    No →     Note in response to ES consultation.

- Does the ES give an accurate account of the policy context against which the proposal and its effects will be considered?

Yes ↓    No →     Note major omissions in response to ES consultation.

- Does the ES properly acknowledge any deficiencies or uncertainties in the information base?

Yes ↓    No →     Note deficiencies in response to ES consultation.

- Does the ES adequately and accurately describe the existing status of natural heritage resources?

Yes ↓    No →     Note errors/omissions in response to ES consultation.

*Complete the table below as fully as possible but concentrate on the important effects.*

| Effects on Natural Heritage | Is it identified in ES? | Proposed Mitigation | Unavoidable, Residual Adverse Effects |
| --- | --- | --- | --- |
| | | | |
| | | | |
| | | | |
| | | | |
| | | | |
| | | | |

Describe any positive enhancement

● Are the predictions of effects clear, comprehensive and reasonable?

Yes ↓    No →    Note concerns in response to ES consultation.

● Have indirect, knock-on and cumulative effects been considered?

Yes ↓    No →    Advise Competent Authority of possible secondary effects, and the need to take account of cumulative effects.

● Will significant effects be avoided or adequately mitigated wherever possible?

Yes ↓    No →    Object to the proposal, unless there are overriding policy reasons in favour of the proposal.

● Are the significant residual adverse impacts of the proposal adequately compensated for?

Yes ↓    No →    Object to the proposal.

● Are there any proposals for enhancement that need to be weighed against the residual adverse impacts of the proposal?

Yes ↓    No →    Object to the proposal if the adverse effects are upon statutory designations.

● Where necessary, has the ES guaranteed the mitigating measures and proposed an effective regime to monitor and redress adverse effects?

Yes ↓    No →    Request that mitigation is guaranteed by conditions and legally binding agreements and that it includes effective monitoring review and remedial or corrective action as may be required.

N.B.  If the ES is revised and resubmitted, fill in Part 2 again, marking the original sheet as 'superseded'. If supplementary information is submitted which changes your views, then amend answers on original sheet indicating that that amendment results from supplementary information.

# Appendix 1 Landscape and Visual Impact Assessment

**For information on the assessment of cumulative landscape and visual impacts of wind energy developments please visit SNH's website: www.snh.org.uk**

## Introduction to this Appendix

**1.** This Appendix explains in more detail the techniques for assessing the landscape and visual impacts of a proposal, within the overall framework of the EA process. Essentially, many proposals are likely to change the landscape and the way in which people see the landscape. The techniques described are based on the current best practice guidance for a systematic approach to landscape and visual impact assessment developed by the Landscape Institute and the Institute of Environmental Assessment with support from SNH. It is set out in the publication *Guidelines for Landscape and Visual Assessment,* The Landscape Institute and the Institute of Environmental Management and Assessment, Spons, 2002, and also *Landscape Character Assessment Guidance for England and Scotland* prepared on behalf of the Countryside Agency and SNH, 2002. Other SNH publications relevant to the assessment of landscape and visual impacts are: (a) *Marine Aquaculture and the Landscape, The Siting and Design of Marine Aquaculture Developments in the Landscape,* (b) *Guidelines on the Environmental Impacts of Windfarms and Small Hydroelectric Schemes* and (c) *Minerals and the Natural Heritage in Scotland's Midland Valley.* A copy of the publication, in hardback book form, should be available to all SNH staff.

> **★ Key advice ★**
>
> If you require further guidance after reading this Appendix, you should refer to the above guidelines and/or your landscape advisors. SNH landscape advisors should be consulted at as early a stage as possible when you are consulted on landscape issues in the scoping stage and on submission of a draft or final Environmental Statement. Consider also the value to you of SNH Landscape Awareness training.

## Introduction to Landscape Planning and the Environmental Assessment Process

**2.** The box overleaf illustrates the key steps in landscape planning. It will be seen that these steps integrate with those of the Environmental Assessment process. For example, looking at alternatives, developing mitigation measures and preparing a detailed assessment for the decision making process.

**3.** In particular, Environmental Assessment for landscape and visual assessments should include:

- Decision on the need for assessment.
- Scoping of the assessment.
- Description of development/proposal.

- Baseline studies.
- Identification of impacts, predict magnitude, durations etc.
- Mitigation.
- Assessment of magnitude, duration etc. of residual impacts.
- Assessment of significance of residual impacts.
- Presentation of findings.
- Consultation.
- Analysis and reporting.
- Decision.

★ Good EIA practice ★

**Appendix 1 Box 1: Key Steps in Landscape Planning**

Understand nature of the landscape.

Identify data, opportunities and constraints.

Modify location, layout, design etc. of all options to achieve best environmental fit.

Prepare strategies to avoid impacts and utilise opportunities.

Compare options, select least harmful.

Develop landscape masterplan.

Prepare landscape and visual impact assessment.

Decision making process.

Detailed design and specification.

Implementation.

After care, maintenance.

Monitoring.

**4.** SNH will mainly be involved in:

- Need for the assessment and scoping.
- Supplying information to baseline studies.
- Advice on mitigation.
- Assessment of residual impacts.
- Consultation.
- Analysis and reporting.

**5.** Your approach to appraisal of landscape and visual impacts will follow the sequence shown in Figure 1 on the next page.

## Appendix 1 Figure 1
### SNH Approach to Landscape and Visual Impact Assessment

Landscape observation and description.

▼

Appreciation of landscape character and landscape change.

▼

Reading about, examining and understanding the proposal –
at various life stages.

▼

Assessing the landscape and visual impacts and their significance.

▼

Considering whether the Environmental Statement is an acceptable basis for the
decision.

▼

Considering whether more or different mitigation is possible and seeking further
information or discussing/negotiating changes.

▼

Drafting a written consultation response.

**6.**     Firstly, however, a general definition of the meaning of landscape and the difference between landscape and visual impacts will provide important background information.

### Definition of 'Landscape'

**7.**     The simplest definition of 'landscape' is 'the appearance of the land'. Landscape is everywhere and may comprise rural landscape, urban landscape (or townscape), urban fringe landscape, coastal landscape, seascape etc.

**8.**     However, human perceptions of place also include things that cannot be seen but which add to the appreciation of places; these are:

● feelings generated by other senses – touch, hearing, smell, taste;
● feelings generated by a knowledge of the place (its cultural and historical associations with people, events, etc.);
● feelings generated by past experience of the place, or similar places – life experience.

**9.**     These combine to give an experience of landscape perceived by all the senses – sight, sound, smell, touch, taste – and by knowledge.

**10.**     What is experienced is influenced by:
● natural and semi-natural features and processes;
● the use and management of the land by humans now;
● the result of the historical use and management of the land;
● cultural associations;
● human activity.

**11.**     SNH takes a comprehensive view of landscape, taking account of more than just the visible components. We recognise that historical and cultural associations and the total experience of landscape through all the senses and through knowledge are integral to understanding landscape character.

**12.** SNH believes that all landscapes, everywhere, are important as:
- an essential part of our natural heritage resource base;
- a reservoir of archaeological and historical evidence;
- an environment for plants and animals, the condition of which directly affects biodiversity conservation;
- a resource that evokes sensual, cultural and spiritual responses essential to human well being;
- an important part of our quality of life, not least as the habitat/environment in which we live.

**13.** SNH recognises that the landscape of Scotland is the direct product of the interaction of innumerable and often extremely complex natural and human influences over thousands of years. The landscape is dynamic and continues to change as a result of natural systems and processes and human influences–land use and management continue to change the components of landscape. The range and scale and speed of change have all increased with technological progress. Armed with modern technology we are able to pay less regard to natural influences–geology, topography, climate, coastal processes – than we had to in the past. This can erode landscape character and local distinctiveness by departing from traditional and more sensitive ways of building and utilising the land that respected natural constraints and used natural, locally available materials.

**14.** Change, however, is inherent in all landscapes. SNH's approach is to manage change, not protect the status quo. SNH believes that a better understanding of landscape, its evolution, management, conservation, restoration and enhancement is essential to achieve environmental sustainability. To reach an improved understanding we need to better appreciate the composition and distribution of landscape types in Scotland, their evolution, the pressures for change that they experience, the likely effects of change and how change may be managed and controlled. The Environmental Assessment process is an important contribution to improving and informing decisions that may affect landscape and visual amenity. The national programme of Landscape Character Assessments is also an important contribution and SNH's responses to Environmental Assessment should be built upon the foundations provided by the local Landscape Character Assessments.

**15.** Environmental Assessment is about the appraisal of **components** of the landscape, appreciating the **character or distinctiveness** of landscape and how changes may affect all of these things. It is not about how individuals may **respond** to the landscape. People's responses to the landscape will vary as a result of their own personal aesthetic taste, tolerance of sound, preferences for smells and tastes, life experiences, philosophies, interests, education and knowledge. Environmental Assessment should not try to consider people's **responses** to landscapes. One person's landscape of wild beauty and tranquillity is another person's landscape of featureless desolation. Environmental Assessment should look at the physical aspects of the landscape and what is experienced but should not attempt to describe or assess people's reactions to these.

## Landscape and Visual Impacts

**16.** Landscape and visual impacts are related but separate, different concepts.

**Landscape Impacts** are on the fabric, character and quality of the landscape.

They are concerned with:

Landscape components

Landscape character – regional and local distinctiveness

Special interests e.g. designations, conservation sites, cultural associations.

**Visual Impacts** are the effects on people of the changes in available views through intrusion or obstruction and whether important opportunities to enjoy views may be improved or reduced.

**17.** Landscape and visual impacts do not necessarily coincide. Landscape impacts can occur in the absence of visual impacts, for instance where a development is wholly screened from available views, but nonetheless results in a loss of landscape elements, and landscape character within the site boundary. Similarly, some developments, such as a new communications mast in an industrial area, may have significant visual impacts, but insignificant landscape impacts. However, such cases are very much the exception, and for most developments both landscape and visual impacts will need to be assessed.

## Landscape Observation and Description: Components of the Landscape

**18.** The components of landscape and the influences on those components are fundamental to our appreciation of landscape character and its distinctiveness. Some of these components are objective, some are subjective. Landscape observation, description and appreciation always involves objective and subjective matters but you can embrace the subjective elements with confidence by confining description to the components of the landscape and not your responses to these components.

**19.** The components of the landscape are its features and characteristics. The landscape includes:
- **visible, physical**, objective, tangible components, e.g. landform, buildings.
- **visible, spatial (rather than physical)**, subjective, intangible components, e.g. scale, pattern, colour, texture etc.
- **non-visible** components that cannot be seen, e.g. sound and cultural associations.

**20.** In order to structure your approach to observation and description, it is useful to have a fieldsheet that acts as an aide-memoire. No standard fieldsheet could be devised that would be appropriate to all the landscape types in Scotland. Example Fieldsheets 1, 2 and 3 at the end of this Appendix, entitled Landscape Observation and Description, are designed to indicate the wide range of features and characteristics that you may find in Scotland; they are certainly not exhaustive. You could, however, modify them to include things relevant to your area and to delete irrelevant ones. You could use your local version in your everyday work.

### Physical Features and Characteristics

**21.** The physical features and characteristics can be grouped under four broad headings or categories (see Example Fieldsheets 1 and 2).

**Appendix 1 Box 2: The Physical Components of Landscape**

Landform (see Example Fieldsheet 1)

Land Cover and Land Use (see Example Fieldsheets 1 and 2)

Linear Features (see Example Fieldsheet 2)

Single Point Features (see Example Fieldsheet 2)

**22.**   These broad categories can be subdivided (See Example Fieldsheets 1 and 2). For example:

Land Cover and Land Use divided into:
    water;
    forestry, woodland and trees;
    agriculture, fields and boundaries;
    settlements;
    other land uses.

**23.**   All of these components are:
real, physical, measurable, tangible–touchable as well as visible.

They can, therefore, be described with total objectivity: a matter of fact, not opinion. **We are not describing our responses to them, e.g. whether we like them or not, just whether they are there or not.** Together they create compositions in infinitely variable ways.

**24.**   Some components will be more significant than others. The significant ones may contribute to the character of the landscape or may form conspicuous features within the landscape that are not typical. In completing the fieldsheet you might develop a system, e.g. of boxes or highlighting, to indicate the most significant, i.e. visually prominent or frequent features.

**25.**   We are not making judgements about good or bad compositions or intrusive features. It is a matter of fact how these components combine and whether particular components occur uniquely or frequently.

## Components of Landscape Experience

**26.**   Sometimes referred to as *'Experiential Characteristics'* and set out on Example Fieldsheet 3 at the end of this Appendix. These are not physical components but may include:

● visible, spatial characteristics that cannot be touched but can be seen (e.g. colour or pattern);
● characteristics that relate to our other senses, such as hearing, smell, taste (e.g. sounds and scents);
● characteristics that are introduced by knowledge of the area (e.g. associations with people, events or cultural heritage or artistic or literary works).

They are all included in the list of components in Box 3 and on Example Fieldsheet 3.

**Appendix 1 Box 3: The Components of Landscape Experience**

**Visible**        Balance, colour, diversity, form, line, management, movement, openness, scale, texture.

**Other Senses** Sound, taste, smell

**Knowledge**   Historical associations, cultural associations (but factual things, not emotional things).

**27.**    In turn, each of the visible components can be described in relative terms. They do not lend themselves to accurate measurement, like the physical characteristics, but they can be described within a range of common adjectives associated with the subject. For example: openness may be described as: tightly enclosed, confined, open or exposed. (See Example Fieldsheet 3.) These adjectives give us a fairly descriptive picture. See other descriptions on Example Fieldsheet 3 at the end of this Appendix.

**28.**    These descriptions are subjective but, nevertheless, meaningful. The likelihood is that most people would describe a component in a particular landscape **in the context of its location in Scotland**, by using the same adjective. Context of location is important. What is open and large scale in the Western Highlands will be different from open, large scale landscapes in the Midland Valley of Scotland. Your description may vary depending on where you are working.

**29.**    These descriptions, then, are capable of portraying a picture of the landscape character, in combination with all the other component descriptions.

**30.**    These descriptions do not relate to our responses to the landscape but our experience of it. If you approach descriptive methods in the right way, your understanding, expression and appreciation of the landscape is valid, you are capable of doing it and there is nothing wrong with subjectivity if it is founded on an informed understanding and structured approach.

**31.**    It is also important to realise that because these components are capable of meaningful description they can also change if the landscape changes. Furthermore, most are capable of being changed by human activity, such as changes in land use or management or development.

**32.**    For example, removing field boundaries will change the scale and openness. Mineral operations may change texture, colour, scale, balance, form, line, movement and sound.

**33.**    These must, therefore, be important components in landscape character and need to be considered in landscape assessment.

## Appreciation of Landscape Character and Landscape Change

**34.**    The components of the landscape combine to create special combinations that everyone sees and feels, no matter what their response to it may be. The combinations of components are more than the sum of their component parts. Landscape character is the combination of all the components.

## Landscape Distinctiveness

**35.**    The combinations of landscape characteristics vary considerably, indeed infinitely, from place to place and usually provide such a unique combination of components that it is distinctive – not quite like anywhere else. This gives a sense of place and identity unique to each area (except for example a monotonous housing estate or forest plantation that is anonymous – it could be anywhere).

## Landscape Character

**36.**    Despite this unique combination of components locally, however, most areas have key components – features or characteristics – that create broad types of combinations that are repeated, or at least occur in more than one area. These broad combinations are identified as **'Landscape Types'**. Their local variations are identified as **'Landscape Character Areas'** or sub-areas. A sense of place for local people comes from their recognition and familiarity with their local area which provides, for them, a strong sense of place and identity even if it is not familiar to other people.

## Landscape Change

**37**    Landscapes are dynamic. They change through natural processes – e.g. maturity of woodlands – and natural systems – e.g. coastal accretion, river erosion. Most changes, however, are the result of human activity, land use, management or neglect.

**38.**    Change is inevitable and can alter the landscape character, making it more or less typical of its landscape type or even changing it to another landscape type altogether. Change in itself is not, therefore, necessarily a bad thing. It can restore or enhance landscape character. Alternatively, it can damage, degrade or destroy landscape character. SNH seeks to manage change, not to prevent it; that would be unrealistic.

**39.    Appreciation of landscape character – what is significant, what is important–is fundamental to landscape planning and management. When considering proposals for change we need to focus on those aspects that form the key components of the landscape and assess the changes to them that would occur:**

**(a) anyway, as a result of trends and natural changes; and**

**(b) as a result of the proposal that is subject to the Environmental Assessment.**

**Reading about, examining and understanding the proposal–at various life stages.**

**40.**    Landscape and visual impacts can arise from a variety of sources. They can be caused by changes in land use, for example mineral extraction, afforestation and land drainage; by the development of buildings and structures such as power stations, industrial estates, roads and housing developments; by changes in land management, such as intensification of agricultural use, which can be a vehicle for biological and landscape change; and, less commonly, by changes in production processes and emissions, for instance from quarries, chemical, food and textile industry plants.

**41.** In order to predict the changes that would result from a project it is necessary to fully understand the project itself. There will be relatively obvious points to familiarise yourself with, such as the location and size or scale of the development and the nature of the project–what it would look like and sound like. There will also be less obvious points to consider, such as the different stages that a project may go through. Reference should be made to the project life cycle at Figure 4 of the main Handbook.

**42.** Means of access or of importing or exporting materials, or energy transmission, water supply etc. could all have landscape and visual impacts including indirect and off site impacts. The excavation of local borrow pits for construction materials, temporary or permanent disposal or storage of waste, topsoil, subsoil, other overburdens and surface water or settlement lagoons could create new features in the landscape.

**43.** The project may necessarily need ancillary or related forms of development which have not been clearly identified and described in the proposal such as: construction yards or compounds; ancillary buildings or structures; jetties; lighting; security fencing; gantries, poles, masts, cranes or towers; signs and even sirens or other audible warning devices.

**44.** The proposal may well contain some mitigation measures which are already incorporated into the scheme. What form do they take, what would be their scale, duration, location and how would they be constructed or implemented? To what extent do they appear to be effective and would they have landscape or visual impacts themselves?

**45.** It takes time to build up this picture of what is proposed, but this is essential before visiting the site and beginning to examine the existing landscape character and views and assessing how they may be affected by the project.

## Predicting the Landscape and Visual Impacts

See Section C.7 of the main Handbook and Example Fieldsheets 4 and 5).

**46.** Impact occurs when landscape or visual resources are affected. Where we have a proposal for assessment there will be **'receptors'** – things that will be affected, e.g.

- **landscape** that is there now;
- **people** that are there now; and

**'impacts'** – the changes that the landscape and the people would experience.

**47.** Receptors of landscape and visual impact may include physical and natural landscape and biological resources, special interests and groups of viewers. Receptors can be, e.g.:

- Specific landscape components, e.g. shoreline, hill or river.
- Areas of distinctive character.
- Valued landscapes like local beauty spots.
- Historic, designed landscapes.
- People – residents, workers, travellers.

**48.** Reference should be made to the full range of types of impacts shown in Box D.7.3 of the main Handbook. (See also Example Fieldsheets 4 and 5.)

# Assessing the Significance of Landscape and Visual Impacts

See Sections C.8 of the main Handbook and Example Fieldsheets 4 and 5 at the end of this Appendix.

**49.** Reference should be made to section C.8 of the main Handbook, which considers the assessment of the significance of impacts. Essentially this depends on, amongst other things:

- the type of impact;
- the magnitude or scale of the impact;
- duration – whether it is a permanent or temporary impact;
- the importance of the receptor as a landscape component (or the number of people affected, what they are doing and the context of the view).

**50.** Significance thresholds can, therefore, be determined from different combinations of sensitivity and magnitude. In order to develop significance thresholds it is necessary first to classify the sensitivity of receptors and the magnitude of change according to reference points along a continuum, as shown in the examples in Figure 2 below. These can be used in your fieldsheets, as in Example Fieldsheets 4 and 5 at the end of this Appendix. You should clearly distinguish between landscape and visual receptors and a useful way of ensuring that you do this is to use separate fieldsheets for landscape receptors and impacts (Example Fieldsheet 4) and visual receptors and impacts (Example Fieldsheet 5).

**51.** In the example in Figure 2 below a scale of 'high, medium and low' has been used, but it must be stressed that this is only an example. Every project will

## Appendix 1 Figure 2
### Examples of Sensitive Receptors and Impact Magnitude Related to Significance of Impacts

| Sensitivity | Significance | Magnitude |
|---|---|---|
| Key features and characteristics of landscape of distinctive character, susceptible to relatively small changes. NSAs, AGLVs | High | Noticeable change in characteristics or features over an extensive area ranging to intensive change to more limited area |
| Moderately significant features and characteristics in a distinctive landscape or a landscape of moderately distinctive character reasonably tolerant of changes | Medium | Moderate or localised changes |
| Unimportant features or characteristics or indistinct landscape character types potentially tolerant of substantial change | Low | Virtually imperceptible changes or changes within the capacity of the landscape to absorb |

### Classification of Sensitive Visual Receptors and Impact Magnitude

| Sensitivity | Significance | Magnitude |
|---|---|---|
| Residential properties, tourist hotels, public rights of way, country parks, viewpoints etc. | High | Majority of viewers affected, major change in view |
| Schools, sporting or recreational facilities not related to enjoyment of the natural heritage | Medium | Many/some viewers affected, moderate change in view |
| Industrial, office or other workplaces | Low | Few viewers affected, minor changes in view |

require its own set of criteria and thresholds, tailored to suit local conditions and circumstances, and it should be remembered that impacts can be positive as well as negative. The benefit of such a system, though, is to help separate fact from interpretation, and hence to simplify discussion and agreement on the significance of impacts. The Example Fieldsheets at the end of this Appendix use a four point 'high/medium/low/insignificant' scale, again to illustrate different approaches that may be applicable in different circumstances.

**52.**  Numerical scoring or weighting should be avoided. Attempting to attach precise numerical values to qualitative resources is rarely successful, and should not be used as a substitute for reasoned professional judgement.

**53.**  As with landscape description it may help to use a fieldsheet or checklist (again modified to your area and or your work) to structure your approach. This helps considerably in drafting the text of your response. When in the field, try to envisage the landscape with the development in place – add and subtract relevant features and consider what effect that would have.

**54.**  Landscape impacts in the checklist may usefully be grouped under 'receptors' with a similar list to those used to describe the landscape components. Thus, you will be using a basis for assessing landscape and visual impact significance directly drawn from your landscape description and related to the key characteristics and features that you identified in your observations (assisted by the Landscape Character Assessment for the relevant area where available). This provides a rational and well reasoned justification for your representations.

**55.**  For each impact you can indicate whether there would be a high, medium, low or insignificant adverse or beneficial effect. If these are related to the significance of the landscape components, in terms of the contribution they make to the character and distinctiveness of the landscape, then you will begin to understand the significance of the impacts.

**56.**  Similarly you can use a fieldsheet/checklist for assessing the visual impacts.

**57.**  You will be dealing with residual impacts – taking mitigation into account but remembering that some mitigation will take time (screen planting) and some mitigation measures can have impacts themselves, e.g. screen mounds can obstruct views and look out of scale and place because of their size and shape.

## Considering whether the Environmental Statement is an Acceptable Basis for the Decision

See Sections D.6 and D.9 of the main Handbook.

**58.**  The next task is to consider whether the Environmental Statement is an acceptable basis for informing the decision maker. This will include checking to see whether the good practice methods described in this Appendix and the LI/IEA Guidance have been adopted, or whether some similarly rational and impartial method has been used that is clearly explained.

**A key test is whether the Environmental Statement clearly distinguishes between landscape and visual impacts; many do not.**

**Does the Environmental Statement fully and fairly describe all relevant and**

**significant landscape and visual impacts and does it assign much the same levels of significance to these impacts as your assessment?**

**59.**   If there are discrepancies or gaps, what seems to be the difference between the Environmental Statement's conclusions and your own, and how might this have arisen?

**60.**   It is not feasible to produce a comprehensive checklist of all the points that you may need to consider when appraising the adequacy and effectiveness of Environmental Statements, owing to the considerable scope of content, project types and methods of presentation. However, some of the points in Box 4 below will usually be worth considering.

★ **Good EIA practice** ★

**Appendix 1 Box 4**
**Useful Tests to Apply to Environmental Statements in Respect of Landscape and Visual Impact Assessments**

Does the Environmental Statement contain what you consider to be fair/accurate/appropriate illustrations?

Is there a Zone of Visual Influence Map or similar cartographic expression of the area affected?

Are there before and after illustrations such as artist's impressions, sketches, photomontage or computer aided montages or overlays?

Are viewpoints fair and typical and comprehensive of relevant views?

Are maps, diagrams and illustrations clear and is the text clear and unambiguous?

Are options or alternatives adequately considered?

Are mitigation measures adequately described and are their effects assessed?

Are residual effects clearly identified and if so could they be further reduced even at costs that the developer may be seeking to avoid?

**Considering whether more or different mitigation is possible and seeking further information or discussing/negotiating changes**

See Sections C.9, D.6, D.7, D.8 and E.4 of the main Handbook.

**61.**   If you consider that important information, which could affect the outcome of the application for which the Environmental Statement has been prepared, is absent or inadequate you should inform the Competent Authority as soon as possible. You may need to consider, in certain circumstances, the use of a holding objection (see SNH Local Authorities Handbook, section D.18 and Appendix IV). In any event you should ask the Competent Authority to require the applicant to submit the information, if necessary as a supplementary Environmental Statement, and ask the Authority not to determine the application until all of the necessary environmental information is available (see main text in Section F.4 of this Handbook). Submission of the required information may mean that you have to reassess the landscape and visual impacts of the proposal.

**62.**   If you conclude that more or different mitigation would be appropriate, or adverse effects could be avoided, or compensated, or new benefits could be achieved (see Annexe 1 of this Handbook for definitions), you should check these matters with your landscape advisor before you consider whether to open negotiations with the Competent Authority and/or the developer to affect changes to the proposals. Whether you do so in advance of, or simultaneously with, the submission of your representations will depend largely on time scales available, previous dialogue, confidence in the authority, likelihood of success and other local circumstances.

**63.    However, procedurally, remember that your response is required primarily on whether the project should be consented or authorised and, if so, on what terms and conditions and if not, why not. You should not risk SNH's views being too late to influence the decision merely because you are awaiting a response to suggested changes.**

**64.**    For planning **applications**, generally follow the procedures set out in the SNH Local Authorities Handbook. In all cases, remain aware of deadlines but enter constructive negotiations wherever it would be advantageous to do so.

# Landscape Observation and Description

## Example Fieldsheet 1

### Location

Viewpoint                                        Date

### Visible, physical components of landform, its features and characteristics

| | | | | |
|---|---|---|---|---|
| High plateau | Peak | Knoll ridge | Spur crags | Outcrops |
| Corrie/gulley | Low plateau | Distinct hills | Rolling hills/slopes | Glen valley |
| Gorge | Bench/terrace | Flats | Wide basin | Confined basin |
| Den | Hollows | Plain | Mounds/moraines | Cliff |
| Coastal brae | Bay | Headland | Beach | Intertidal |

Notes

### Land cover and land use – water

| | | | | |
|---|---|---|---|---|
| Sea | Sea loch | Intertidal | Mud/sand | Delta |
| Estuary | Loch | Lochans | Pools | River |
| Whitewater | Burn | Drain/ditch | Canal | Waterfall |
| Reservoir | | | | |

Notes

### Land cover and land use – forestry, woodland and trees

| | | | |
|---|---|---|---|
| Coniferous plantation | Mixed plantation | Broadleaved plantation | Semi-natural woodland |
| Tree clumps/copses | Shelterbelts/tree lines | Roadside tree belts | Policy/parkland trees |
| Hedgerow trees | Notable single trees | | |

Notes

### Land cover and land use – agriculture

| | | | |
|---|---|---|---|
| Arable | Horticulture | Intensive livestock | Ley grassland |
| Permanent pasture | Unimproved grassland | Rough hill grazing | Moorland |
| Animals: | Cattle | Sheep | Pigs |
| | Poultry | Horses | Deer |

Notes

### Land cover and land use – fields and boundaries

| | | | |
|---|---|---|---|
| Stone dykes | Dykes with fencing | Remnant dykes | Continuous hedgerows |
| Hedgerows with gaps | Remnant hedgerows | Lost hedgerows | Post and wire fencing |
| Post and rail fencing | High stone walls | Stone pillars | Wooden/metal gates |
| Beech hedges | Hawthorn hedges | | |
| Field size: | Very Large | Large | Medium | Small |

Notes

# Landscape Observation and Description

## Example Fieldsheet 2

### Location

Viewpoint                                              Date

### Land cover and land use – other land uses

| | | | | |
|---|---|---|---|---|
| Country park | Urban park | Nature reserve | Car parks | Sports fields |
| Golf course | Angling | Camping site | Caravan site | Marina/boats |
| Dock/harbour | Military | Open cast coal | Sand and gravel | Hard rock |
| Industrial | Warehousing | Airfield | Retail | Utilities |

Notes

### Land cover and land use – settlements

| | | | | |
|---|---|---|---|---|
| Nucleated | Scattered | Linear | Unplanned | Model/planned |
| Traditional | Modern | Mixed | Frequent | Infrequent |
| Absent | Town | Village/township | Hamlet | Sprawling |
| Steadings | Regular | Irregular | Absent | Frequent |
| | Infrequent | Small | Medium | Large |
| | Traditional | Modified | Extended | Converted |

### Dominant Building Materials

| | | |
|---|---|---|
| Stone colour | Brick colour | Render/colourwash |
| Tile roof colour | Slate roof colour | Stone roof colour |

Notes

### Linear features

| | | | | |
|---|---|---|---|---|
| Motorway | Main road | B roads | Minor roads | Tracks |
| Bridleways/paths | Drove roads | Hill tracks | Derelict/operational railway | |
| Embankments | Cuttings | Powerlines | High voltage | Low voltage |
| Rivers/watercourses | Overhead telephone | Pipelines | Coast/shoreline | |

### Single point features

| | | | | |
|---|---|---|---|---|
| Church | Castle | Ruin | Folly | Obelisk |
| Cairn | Bridge | Large house | Steadings | Signs |
| Mast/transmitter | Industrial site | Tips/bings | Quarry/mine | Quarry buildings |

Notes

# Landscape Observation and Description

## Example Fieldsheet 3

### Location

| Viewpoint | Date |
|---|---|

### Components of landscape experience–visible/spatial characteristics

| SCALE | Intimate | Small | Large | Vast |
|---|---|---|---|---|
| OPENNESS | Tightly enclosed | Confined | Open | Exposed |
| COLOUR | Monochrome | Muted | Colourful | Garish |
| TEXTURE | Smooth | Varied texture | Rough | Craggy |
| DIVERSITY | Uniform | Simple | Diverse | Complex |
| FORM | Vertical | Sloping | Rolling | Flat/horizontal |
| LINE | Straight | Angular | Curved | Sinuous |
| BALANCE | Harmonious | Balanced | Discordant | Chaotic |
| MOVEMENT | Dead | Calm | Active | Busy |
| PATTERN | Random Indistinct | Organised Irregular | Planned Regular | Formal Geometric |
| MANAGEMENT | (Semi) Natural | Derelict/disturbed | Tended | Manicured |

### Components of landscape experience–other senses

| SOUND | Silent | Quiet | Disturbed | Noisy |
|---|---|---|---|---|
| SMELL | Fresh | Agricultural | Coastal | Industrial |
| OTHER | | | | |

| Notes |
|---|

# Landscape Observation and Description

## Example Fieldsheet 4

### Location

Viewpoint                                                    Date

### Proposal

| Landscape receptors What will be affected? | Sensitivity How important is it? | Impact What will the effect be? | Significance of impact |
|---|---|---|---|
| Landform | High/Medium/ Low/Insignificant | | High/Medium/ Low/Insignificant |
| Water | High/Medium/ Low/Insignificant | | High/Medium/ Low/Insignificant |
| Woodland and trees | High/Medium/ Low/Insignificant | | High/Medium/ Low/Insignificant |
| Agriculture | High/Medium/ Low/Insignificant | | High/Medium/ Low/Insignificant |
| Fields and boundaries | High/Medium/ Low/Insignificant | | High/Medium/ Low/Insignificant |
| Other land uses | High/Medium/ Low/Insignificant | | High/Medium/ Low/Insignificant |
| Settlement pattern | High/Medium/ Low/Insignificant | | High/Medium/ Low/Insignificant |
| Linear features | High/Medium/ Low/Insignificant | | High/Medium/ Low/Insignificant |
| Point features | High/Medium/ Low/Insignificant | | High/Medium/ Low/Insignificant |
| Aspects of landscape experience | Key features and characteristics | | High/Medium/ Low/Insignificant |

# Landscape Observation and Description

## Example Fieldsheet 5

### Location

Viewpoint                                                        Date

| Visual receptors | Sensitivity of viewpoint? | Impact Visual intrusion/ obstruction | Significance of impacts |
|---|---|---|---|
| Trunk roads and motorways | High/Medium/Low/ Insignificant | | High/Medium/Low/ Insignificant |
| A and B roads | High/Medium/Low/ Insignificant | | High/Medium/Low/ Insignificant |
| Minor roads | High/Medium/Low/ Insignificant | | High/Medium/Low/ Insignificant |
| Rights of way | High/Medium/Low/ Insignificant | | High/Medium/Low/ Insignificant |
| Important viewpoints | High/Medium/Low/ Insignificant | | High/Medium/Low/ Insignificant |
| Railways | High/Medium/Low/ Insignificant | | High/Medium/Low/ Insignificant |
| Open space and recreation areas | High/Medium/Low/ Insignificant | | High/Medium/Low/ Insignificant |
| Public buildings | High/Medium/Low/ Insignificant | | High/Medium/Low/ Insignificant |
| Residential properties | High/Medium/Low/ Insignificant | | High/Medium/Low/ Insignificant |
| Workplaces | High/Medium/Low/ Insignificant | | High/Medium/Low/ Insignificant |

# Appendix 2   Ecological Impact Assessment

## Introduction to this Appendix

1.     This Appendix explains in more detail the techniques for assessing the ecological impacts of a proposal, within the overall framework of the Environmental Assessment process. The techniques described here are based on current best practice and incorporate the main points made in the publication *Guidelines for Baseline Ecological Assessment* written by the Institute of Environmental Assessment and published by Spons in 1995. A copy of the publication should be available to all SNH staff. If you require further guidance on baseline ecological information after reading this Appendix, you should refer to the book or to your specialist advisors.

## Summary of the Environmental Assessment Process: Ecological Assessment

2.     The Environmental Assessment Process is described in Section B.1 and Figure 1 of the main Handbook. SNH's involvement in ecological assessment will mainly relate to the following steps:

●     The decision whether an Environmental Assessment is required
●     Scoping of the issues to be addressed in the assessment
●     Collection of information and undertaking of surveys
●     Consultation on draft or published Environmental Statement
●     Monitoring the effects of implementation.

## Advising on the Scope of an Ecological Study
See Section C.4 of the main Handbook.

3.     Owing to the wide variation of habitats and of the proposals which may affect them, there can be no standard approach to ecological issues in an Environmental Assessment. Each assessment will be unique and both the methods and the ecological issues under consideration must relate to the particular circumstances of each Environmental Assessment. Correctly defining the scope of ecological work in an Environmental Assessment is essential in securing a good quality Environmental Statement, saving unnecessary expense and helping SNH in their future consideration of the proposal's effects. Work spent on scoping can pay great dividends later in the process. The authors of Environmental Statements should be given every encouragement by SNH to undertake and produce a scoping report and to agree its contents before proceeding with, or certainly before completing, the main Environmental Statement.

4. Your involvement in scoping has 2 major aims:

●     to decide whether the proposal raises issues of ecological importance;
●     where significant impacts may occur, to determine whether there is sufficient ecological information already available to assess the magnitude of these impacts or additional survey information will be required, and if so what is required and how the surveys and assessments will be done.

**5.**   In determining these two points, you should carry out the following tasks:

- review the quality and extent of all existing ecological data collected by the developer for the site and its surrounds, including presence/absence of rare or protected species and recognised sites of nature conservation interest (both statutory and non-statutory);
- consult with colleagues and others with local ecological knowledge;
- visit the site and identify general locations of any current areas of specific interest for floral or faunal communities, and also consider the potential effect of the proposal on the wider ecological framework e.g. desirable ecological features such as watercourses, hedgerows, woodlands, etc. not benefiting from any nature conservation designation.

**6.**   In some cases, the above tasks will not be sufficient to enable decisions to be taken and there may be a need for more information than can be produced by desk study and a site visit. SNH may well have to request, therefore, that some preliminary field survey work is carried out before decisions are made on the full scope of ecological issues in the Statement. In most cases, this preliminary work will consist of producing a Phase 1 Habitat Survey of the site (and often of surrounding land as well), together with target notes highlighting the value of certain habitats for both floral and faunal communities.

**7.**   Once all relevant preliminary information is available, then decisions can be taken on the extent of detailed survey work and the methods to be employed, as discussed below.

## Advising on the Extent of Survey Work
See Sections C.5 and C.6 of the main Handbook.

**8.**   The extent or area to be covered by ecological survey work will vary considerably from case to case. The essential requirement is to consider the context of the proposal with the surrounding area, i.e. the interaction between the two. A pipeline with a total working construction width of 40 metres in an arable landscape will have a much smaller ecological footprint, than say, an estuarine power station. Both proposals may need an Environmental Assessment, but the extent of the area of concern will be very different.

**9.** In considering possible off-site impacts of a proposal, special consideration should be given to the following factors:

- **noise**, particularly its effect on bird populations;
- **hydrological effects**, e.g. effects on the water table, changes in flood patterns, downstream effects;
- **air quality** – proposals such as new power stations can affect air quality over large areas with consequent implications for the natural heritage;
- **effects on badgers, otters, deer and other large mammals** of proposals such as roads, which may present obstacles or hazards to their movements;
- **coastal processes**, which may need consideration over large areas if affected by estuarine or coastal developments.

It will often be necessary to seek guidance at this stage from your specialist advisors.

**10.**   As a general rule, unless one of the above factors applies, it will usually be unnecessary to undertake any ecological survey work outwith a 2 kilometre radius of the development.

## Advising on Survey Methods

See Sections C.5 and C.6 of the Handbook.

**11.** SNH has an important opportunity during scoping and early liaison to influence the choice of survey methods. You should be able to advise on:

- the survey methods proposed by the developer for specific ecological groups;
- the correct time of year for surveying particular species or groups;
- the competencies required to undertake aspects of ecological survey work;
- the choice of criteria to be used in evaluating ecological attributes.

**12.** You must be satisfied that:

- the survey methods to be employed will provide enough information on baseline conditions to enable impacts to be properly assessed;
- field surveyors will comply with the relevant licensing procedures for studying protected species.

**13.** The criteria, suggested by the IEA, for more detailed surveys are summarised for each ecological group in the boxes below. The timing of surveys is also given, but see also Figure 1 below.

### 14. Survey Criteria: Vegetation

Detailed Surveys of Vascular Plants should be undertaken where:

a. the development may affect any plant species:
- listed in the Red Data Book
- in Schedule 8 of the Wildlife and Countryside Act
- in Appendix 1 of the Bern Convention
- in Annexes II and IV of EC Habitats Directive
- listed in *Scarce Plants in Britain*, published by JNCC.
b. there are habitats of statutory significance for vascular plants, especially priority habitats as listed in Annexe I of the EC Habitats Directive;
c. there are vegetation types of potential regional or local (i.e. district) importance.

### Survey Methods for Vascular Plants

a. In most situations, surveys should be in accordance with the National Vegetation Classification (NVC).
b. Where a protected or rare species is the main issue, a survey of species distribution and abundance may be more appropriate than an NVC survey.

N.B. For most ecological assessments only Phase 2 surveys of vascular plants are usually required. However, surveys of lichens and bryophytes should be undertaken where protected, Red Data Book or Nationally Scarce species may be affected, or where habitats recognised as having significant interest for these species may be affected. Further surveys of freshwater algae should be undertaken where there could be an impact on a protected, Red Data Book or Nationally Scarce species of stonewort.

## 15. Survey Criteria: Birds

Detailed Surveys of Birds should be undertaken where:

a. the development may affect a breeding pair or population of a species occurring on Schedule 1 of the Wildlife and Countryside Act or Annexe I of the EC Birds Directive;

b. suitable breeding habitat on the development site is in close proximity to known populations of Schedule 1 or Annexe I species, e.g. pine plantations close to known breeding locations of crossbill;

c. a habitat holding at least 1% of the UK population of a species may be affected;

d. the impact area includes vulnerable habitats associated with scarce breeding birds, e.g. montane areas, native pine woods, machair;

e. there are species or populations of importance at a regional or local (i.e. district) level.

### Survey Methods for Birds

a. Breeding bird surveys, including their timing, must be targeted on the species or habitat of interest.

b. Use of the full Common Bird Census or Waterways Birds Survey, both developed by the BTO, are only of value when the breeding location of all species is important. They are rarely suitable for most scarce species.

c. Surveys of wintering birds should, if possible, present several years' data from existing records. Where this is not possible monthly counts on roosting and feeding sites should be undertaken.

## 16. Survey Criteria: Mammals

Detailed Surveys of Mammals should be undertaken where:

a. the development may affect badgers, seals or species protected under:
- the Bern, Bonn and ASCOBANS conventions
- the Wildlife and Countryside Act
- the EC Habitats Directive;

b. a development presents an obstacle or hazard to the movement of large mammals, e.g. deer or badgers, crossing new roads, or obstructions to otters in rivers;

c. a population of mammals has an important influence on ecosystems in and around a proposed development, e.g. grazing by rabbits;

d. species are not covered by existing legislation, but are reasonably recognised as being of local importance.

### Survey Methods for Mammals

Survey methods for mammals will be different for different species and will vary as a result of the requirements of each study. Reference should be made to established sources of guidance for each species.

The species of mammal most likely to require further study are badger, bat, otter, pine marten, red squirrel and wildcat.

## 17.    Survey Criteria: Amphibians and Reptiles

Detailed Surveys of Amphibians and/or Reptiles should be undertaken where:

a. sites are known to contain:
   - protected species
   - good assemblages of species
   - species at the edge of their geographical range;
b. sites lie within the known geographical range of a particular species and contain suitable habitat for that species.

### Survey Methods for Amphibians

The recommended survey methods for newts are netting, torch counting and bottle trapping. Counts of newts should be undertaken at night between April and the end of July.

Counts of common toad should be carried out in April, during the night for adults, and during the day for spawn strings.

Spawn clumps of common frog should be counted during the day in March/April.

## 18.    Survey Criteria: Fish

Detailed Surveys of Fish should be undertaken where:

a. the development may affect species:
   - listed in Schedule 5 of the Wildlife and Countryside Act
   - listed in Appendix 2 of the Bern Convention
   - listed in Appendix II or V of the EC Habitats Directive;
b. species are known to be in decline in the UK: allis shad, twaite shad, Arctic charr, powan, vendace, pollan, smelt, burbot;
c. development may affect unusual races, for example, 'landlocked' river lamprey in Loch Lomond, the spineless three-spined stickleback in the Hebrides;
d. there are important fish communities, i.e. with unusual assemblages, e.g. Loch Lomond, Loch Eck;
e. a fishery may be affected, for example, by interruption to fish migration, damage to spawning grounds.

### Survey Methods for Fish

Appropriate methods vary greatly according to fish species, life stage and habitat. The description below provides a brief summary of available survey methods:

**Streams** Quantitative electro-fishing is the most accepted method, though drift nets for larvae and traps and counters for migratory species may also provide reliable data.

**Rivers** Electro-fishing is more difficult in larger rivers and here seine netting and tow nets (for larvae) are more effective.

**Ponds** Estimates are best obtained from mark-recapture methods, using fish caught by seine netting or trapping.

**Lakes** In larger lakes, indicative numbers may be feasible using mixed-mesh grill nets and traps. Ichthyoplankton nets and other specialised samples may be required to sample pelagic larvae.

## 19.    Survey Criteria: Terrestrial and Aquatic Invertebrates

Detailed Surveys of Invertebrates should be undertaken where:

a. developments may have an impact on freshwater quality, a baseline survey should be undertaken unless adequate data already exists. If the survey reveals 26 or more families present, or a BMWP score >150, or an ASPT score >6.48, then the sample should be analysed to species level wherever practicable;

b. the Phase 1 survey identifies features or habitats of significant value to invertebrates, e.g. dying timber, ancient woodland and fens;

c. the desk study reveals that the site is a key dragonfly site, as this is a good indicator for quality invertebrate habitat;

Further surveys of terrestrial invertebrates should be undertaken where Red Data Book or Nationally Scarce species may be affected or where habitats similar to nearby areas of known invertebrate interest lie within the impact area.

### Survey Methods for Invertebrates

Surveys for terrestrial insects and most other invertebrates should be carried out between May and September.

Field surveys of terrestrial invertebrates should be restricted initially to a few target groups, e.g. Carabidae, Lepidoptera, Orthoptera, Syrphidae, Odonata. The identity of any Red Data Book species must be confirmed.

Several biotic indices have been developed to assess water quality. The most widely used are the BMWP score and the Environmental Quality Index from the Rivers Invertebrate Prediction and Classification System.

Sampling methods should be standardised to allow comparisons between sampling sites and over time, e.g. standard number of kick samples, net sweeps.

## Appendix 2 Figure 1. Timing of Ecological Surveys

Times of year at which field surveys for various groups of organisms and features are generally best carried out. There are, of course, some exceptions, e.g. hay meadows cannot be surveyed for vascular plants after cutting, sand dunes should be surveyed for spring annuals before the middle of May and breeding activity in some birds, such as owls, can occur in almost any month of the year.

O = Optimal time    S = Sub-optimal Time   P = Poor time   No entry = Unacceptable time

| Groups | Jan | Feb | Mar | April | May | June | July | Aug | Sep | Oct | Nov | Dec |
|---|---|---|---|---|---|---|---|---|---|---|---|---|
| Vascular plants | | | | PP | PPO | OOO | OOO | OOO | OSS | SSS | PPP | PPP |
| Bryophytes, lichens | SSS | SSS | OOO | OOO | SSS | SSS | SSS | SSS | SSS | SSS | SSS | SSS |
| Marine algae | | | PP | PPP | PPO | OOO | OOO | OOO | OSS | SPP | | |
| Large fungi | | | | | | SSS | SSS | SSS | OOO | OOO | SPP | |
| Wintering birds | OOO | OOO | SSS | | | | | | | PP | SSO | OOO |
| Breeding birds | | | SS | OOO | OOO | OOO | | | | | | |
| Lepidoptera | | | PPP | SSS | OOO | OOO | OOO | SSS | SSS | SSS | PP | |
| Dragonflies | | | | | | PSS | SSO | OOO | OOS | PP | | |
| Aquatic insects | SSS | OOO | OOO | SSS | | | | | | PPP | PPP | SSS |
| Ancient woodland features | OOO | OOO | OSS | SS | | | | | | | SSS | OOO |

## Advising on the Prediction and Assessment of Ecological Impacts

See Sections C.7, C.8 and D.9 of the Handbook.

**20.**   Impact occurs when ecological resources are affected. Where there is a proposal for assessment there will be:

**'receptors'** – things that will be affected, e.g. habitats and species that are there now; and

**'impacts'** – the changes that the habitats and species would experience as a result of the development or proposal.

**21.**   Impacts may be beneficial or adverse, direct or indirect, temporary or permanent, single or cumulative, and of course may vary in their duration, timing, magnitude and significance. There may be different impacts at different stages of the project. Reference should be made to Box D.7.3 of the main Handbook for the full list of potential impacts and to Figure 4 for impacts associated with the different life stages of a project.

**Assessment** of the proposal involves:

- identifying the receptors,
- identifying and predicting the impacts (changes), and then
- assessing the significance of the changes so the appraisal may contribute to the decision whether it should be allowed to proceed, modified or prohibited.

**Significance** depends on:

- the importance of the receptor, i.e. the importance of the ecological features, habitats and species present at any given location;
- the timing, magnitude and duration of the impact.

**22.**   Reference should be made to Sections C.3 and D.8 of the main Handbook which address the question of significance. Significance thresholds can be determined from different combinations of sensitivity and magnitude. There is no accepted practice for categorising degrees of significance, but it is good practice for assessors to set out a matrix or scale for determining significance. An example of such a matrix is given in Figure 2 below:

| Matrix showing determination of description of ecological significance | | | | |
|---|---|---|---|---|
| **Magnitude of Impact** | **Importance of Receptor** | | | |
| | **International** | **National** | **Regional** | **Local** |
| Severe | Exceptional | Exceptional | High | Moderate |
| Moderate | Exceptional | High | Moderate | Low |
| Slight | High | Moderate | Low | Negligible |

**23.**   In the matrix in Figure 2 impacts could be described as follows:

**Severe:**   Wholesale change of the majority of a site or species population.
**Moderate:**   Substantial but partial change to a site or species population.
**Slight:**   Minor change to part of a site or species population, or loss of a very small proportion of a site or population.

**24.**   Every project will require its own set of criteria and terms, tailored to suit local conditions and circumstances. It should be remembered that impacts can be positive as well as negative and all should be addressed impartially.

**25.**   The relative importance of the various ecological receptors should emerge from the baseline description and evaluation of the impact area. Ecological assessors should be encouraged to draw up a checklist showing the relative importance of ecological elements so that the evaluation is explicit and open to reasoned challenge. Impacts on each receptor can then be described and an appropriate term selected to summarise the degree of significance of the anticipated change.

**26.**   Wherever possible, factual information should be given, either in absolute terms or as a percentage of habitat area or species population, e.g.:

- four ponds totalling 0.24 ha would be lost to the scheme
- 5645 m of ditches and streams would be lost or culverted
- 2495 m of new hedgerow would be established.

**27.**   Ideally, to ascertain the true significance of a proposal's effects, the **'do-nothing comparison'** should be considered. The do-nothing alternative considers how the site would change if the proposal did not receive consent. It examines current trends, including the likely level of site management to form a reasoned conclusion about the site's future without the proposal. However, you should be wary that this approach is not misused to describe a worst case scenario.

## Advising on Mitigating Measures

See Section C.9 and D.7 of the Handbook.

**28.**   One of the main aims of Environmental Assessment is to avoid significant adverse effects. However, if a proposal is to go ahead, it will not always be possible to avoid effects, although there will usually be opportunities to reduce or minimise adverse impacts by the use of mitigating measures, such as:

● locating project elements to reduce adverse effects;
● using construction and operation methods which reduce adverse effects, e.g. to avoid disturbance at critical times of the year;
● introducing specific measures into project design, that will reduce adverse effects, e.g. including silt traps in new drains to control pollution from surface water run-off.

**29.**   Ecological mitigation usually focuses on attempts to minimise habitat loss and effect on site integrity, or to minimise disturbance to a habitat or species found within it. Techniques to mitigate short-term disruption depend upon the presence of similar habitats nearby and the likely success of recolonisation and recovery. Habitat or species translocation may have a role in mitigating for adverse effects. However, such techniques are often of uncertain effectiveness and should only be considered as a last resort.

**30.**   The effectiveness of mitigating measures should be addressed in the Environmental Statement. Indeed, the ecological effects of mitigating measures themselves should also be assessed. Measures are often added at a late stage, perhaps to reduce noise or visual intrusion, but such measures could, for example, lead to further habitat loss or the obstruction of wildlife routes.

## Considering the Environmental Statement

See sections D.9 and D.10 of the main Handbook.

31.   A primary task for SNH will be to consider whether an Environmental Statement is an acceptable basis for informing the decision maker. This will include checking whether the good practice methods described in this Appendix and the IEA Guidance have been adopted. You will need to consider a number of issues for each Environmental Statement, and the range of these issues will vary depending on the project type and the approach and presentation adopted in each Environmental Statement. Use the Review Package in Appendix 6 of this Handbook to guide your assessment, but it may be helpful at the outset to consider the effectiveness of an Environmental Statement in respect of the following heads:

● **Description:** is the proposal clearly described?
● **Scope:** does the Environmental Statement properly address all relevant ecological issues?
● **Information:** is the ecological information provided reasonably up-to-date and adequate for assessment purposes?
● **Evaluation:** has the ecological value of sites/species been properly described or evaluated?
● **Prediction:** have all important impacts been identified? Do you agree with the judgements made about the significance of these impacts?
● **Mitigation:** have all possible mitigation measures been explored?
● **Monitoring:** if necessary, are mechanisms proposed which can monitor effects on sensitive receptors and trigger remedial action?

- **Commitment:** what provisions are (or should be) in place to ensure mitigation/monitoring is carried out?

## Seeking further information or discussing/negotiating changes

See Section D.6 to D.8 of the main Handbook.

**32.** If you consider that important information, which could affect the outcome of the application for which the Environmental Statement has been prepared, is absent or inadequate you should inform the Competent Authority as soon as possible. You may need to consider, in certain circumstances, the use of a holding objection (see SNH Local Authorities Handbook, section D.18). In any event you should ask the Competent Authority to require the applicant to submit the information, if necessary as a supplementary Environmental Statement, and ask the authority not to determine the application until all of the necessary environmental information is available (see main text of this Handbook, sections E6, E7 and E.8). Submission of the required information may mean that you have to re-assess the ecological impacts of the proposal.

**33.** If you conclude that more or different mitigation would be appropriate, or adverse effects could be avoided, or compensated, or new benefits could be achieved (see D.9 of this Handbook), then you should consider whether to open negotiations with the Competent Authority and/or the developer to effect changes to the proposals. Whether you do so in advance of, or simultaneously with, the submission of your representations will depend largely on time scales available, previous dialogue, confidence in the authority, likelihood of success and other local circumstances.

**34. However, procedurally, remember that your response is required primarily on whether the project should be consented or authorised and, if so, on what terms and conditions and if not, why not. You should not risk SNH's views being too late to influence the decision merely because you are awaiting a response to suggested changes.**

**35.** For planning applications, generally follow the procedures set out in the SNH Local Authorities Handbook. Remain aware of deadlines but enter constructive negotiations wherever it would be advantageous to do so.

## Drafting a written consultation response

See Section D.10 of the Handbook.

**36.** Follow the guidance in the main Handbook at Section E.10 and guidance in the Local Authorities Handbook at D.18 and Appendix V. Your response should clearly lead with SNH's representations in respect of the application which the Environmental Statement accompanies or relates to. The representations should clearly indicate whether SNH considers that the project may be consented and if so, subject to what conditions or restrictions SNH consider to be appropriate. Draw upon the Environmental Statement and your own assessment to support or justify or argue your case. If you praise or criticise the Environmental Statement ensure that such comments are relevant to your overall representations about the application and refer to relevant issues.

# Appendix 3   Earth Heritage Impact Assessment

## Introduction

**1.**   This Appendix explains in more detail the issues likely to arise in the Environmental Assessment process in respect of earth heritage conservation. There are no published Environmental Assessment techniques or good practice methods relating specifically to earth heritage impact assessment. Earth heritage issues are often overlooked in published Environmental Statements and, unless a geological or geomorphological SSSI is involved, Competent Authorities may also overlook potential earth heritage impacts. Consequently, consideration of these impacts may be absent or inadequate at any stage in the Environmental Assessment process and one of the key objectives of the guidance in this Appendix is to enable SNH to remedy such deficiencies.

**2.**   The Environmental Assessment process described in the main text of this Handbook is entirely relevant and applicable to earth heritage impact assessment. Equally, earth heritage issues should be an integral consideration at every step in the process. This Appendix:

a. sets out the importance of earth heritage considerations in Environmental Assessment;
b. summarises the general classification of earth heritage sites and their conservation objectives relevant to the Environmental Assessment process;
c. identifies the main or typical threats to earth heritage conservation, i.e. the main potential impacts, and project types particularly relevant to earth heritage conservation; and
d. provides general advice on assessing the significance of earth heritage impacts.

### References

**3.**   Reference is made here to the following publications:
Nature Conservancy Council, 1990/1991, *Earth Science Conservation in Great Britain: A Strategy, and Appendices: A Handbook of Earth Science Conservation Techniques*;
Wilson RCL (ed.), 1994, *Earth Heritage Conservation*. The Geological Society in association with The Open University, Milton Keynes;
SNH information and advisory notes on earth heritage conservation.

### Importance of Earth Heritage Considerations in Environmental Assessment

See Sections B.4 and C.3 of the main Handbook.

**4.**   Earth heritage considerations are an essential element of the Environmental Assessment process and any significant impacts on earth heritage features and sites must be included in an Environmental Statement.

**5.** Annexe III of the Environmental Assessment Directive and Schedule 4 of the

EIASR 99 require that an Environmental Statement must include a description of the aspects of the *'environment likely to be significantly affected by the development, including, inter alia, landscape, soil and water and the interrelationship between them and all other aspects of the environment.'*

**6.** Where significant adverse effects are identified the Environmental Statement must include a description of mitigation measures.

**7.** Schedule 4(1) of the EIASR 99 also specifies that an Environmental Statement may include, by way of explanation or amplification, information on, inter alia:
b. the nature and quality of materials to be used in production processes;
c. the type and quantity of expected pollutants including pollution of soils and water;
d. the likely significant direct and indirect effects of the proposal which may result from the use of natural resources including secondary, cumulative, short, medium and long-term, permanent, temporary, positive and negative effects.

**8.** Regulations 19, 36 and 60 provide planning authorities and the Secretary of State with the power to require the above information (and any other information in Schedule 4), having regard in particular to current knowledge and methods of assessment, where it is reasonably required to give proper consideration to the likely environmental effects of the proposed development.

**9.** Thus, all earth heritage interests can and should be included in an Environmental Statement and throughout the Environmental Assessment process wherever the effects of a proposal are likely to be significant. Where they are not included SNH should normally be able to request the Competent Authority to require the developer to submit the information before they grant any consent for the project. In case the Competent Authority disagrees with this approach, the precise references in the Regulations which can be used to press for earth heritage matters to be included in the Environmental Statement are listed below.

## Earth Heritage References in the Regulations

| Landform | Landscape | Schedule 4(3) |
| --- | --- | --- |
| Geological exposures/features | Landscape | Schedule 4(3) |
| River systems | Landscape | Schedule 4(3) |
| | Water | Schedule 4(3) |
| | Soil | Schedule 4(3) |
| Coastal processes | Landscape | Schedule 4(3) |
| | Water | Schedule 4(3) |
| | Soil | Schedule 4(3) |
| Minerals | Natural resources | Schedule 4(4)(b) |
| Soil | Soil | Schedule 4(3) |
| | Natural resources | Schedule 4(4)(b) |

## Earth Heritage Site Classification and Objectives Relevant to Environmental Assessment

**10.** The potential effects of a project on earth heritage interests will usually depend on 2 main considerations:

a. the type of earth heritage site or feature; and

b. the type of project, including its nature, scale, location, duration etc.

**11.** Impact assessment therefore needs to take account of the differing issues and conservation objectives for earth heritage sites. Table 1 below summarises the classification of earth heritage sites and indicates the changing emphasis of the key conservation objectives.

## Types of Impact
See Sections C.4, C.7, C.8 and C.9 of the main Handbook.

**12.** All likely significant effects on earth heritage interests should be assessed. Generally, effects, or impacts, are likely to fall into one or more of the categories summarised in Table 2 below. For each category, examples of potential impacts are given.

### Appendix 3 Table 1
### Site Classification and Conservation Objectives Relevant to Environmental Assessment

| Classification | Site Types | Conservation Objectives |
|---|---|---|
| Integrity Sites | Coastal cells | Minimise changes, avoid significant interference with natural processes and preserve integrity of physical attributes, composition, structure and visibility of systems and sites. |
| | River systems | |
| | Other active geomorphological areas/sites | |
| | Caves and karst sites | |
| | Static geomorphological sites, e.g. kames, eskers | ▲ |
| | Unique mineral or fossil sites | ▲ |
| | Old mine dumps/bings | ▲ |
| Exposure Sites | Inland natural outcrops | ▼ |
| | Stream sections | ▼ |
| | Exposures in disused quarries | ▼ |
| | Stratotype or other key exposures in coastal cliffs or foreshore | Preserve exposures judging changes on their merits in terms of exposure, and where required, enhance the sites. |
| | Exposures in active quarries | |

## Appendix 3 Table 2 Potential Earth Heritage Impacts

| Indirect /Direct | Type | Example | Timescale | Reversibility | Comments |
|---|---|---|---|---|---|
| Direct | Loss | Landtake from site or feature | Permanent | Irreversible | Usually adverse can be cumulative |
| | Removal | Mineral extraction from geological feature e.g. a kame | Permanent | Irreversible | Usually adverse can be cumulative |
| | Fragmentation | Partial removal of features | Permanent | Irreversible | Usually adverse, often cumulative |
| | Burial | Burial by landfill of quarry or cutting | Permanent | Irreversible | Usually adverse |
| | Obscuring/ covering | Afforestation over geological features | Long-term | Reversible | Usually adverse can be cumulative |
| | | Mineral overburden dump on geological features | Medium-term | Reversible | Usually adverse |
| | | Screen mounds around construction site | Short-term | Reversible | Usually adverse |
| | Changes to natural systems | River engineering works/ flood defences | Permanent or long-term | May be irreversible | Usually adverse can be cumulative |
| | Changes to coastal processes | Coast protection works | Permanent or long-term | May be irreversible | Usually adverse can be cumulative |
| Indirect | Consumption of natural resources | Mineral extraction | Permanent | Irreversible | Usually adverse |
| | Changes to natural systems | River engineering works/ flood defences | Permanent or long-term | May be irreversible | Usually adverse can be cumulative |
| | Changes to coastal processes | Coast protection works | Permanent or long-term | May be irreversible | Usually adverse can be cumulative |
| | Obstructing access | Closure of paths to geological features | Various time scales | Usually reversible | Usually adverse |
| | Enhancing access | Provision of access and/ or interpretation | Various time scales | Usually reversible | Usually beneficial |
| | Obscuring views of geological and landform features | Afforestation | Long-term | Reversible | Usually adverse can be cumulative |
| | Changes to setting and context | Built development | Permanent | Irreversible | Usually adverse can be cumulative |

## Project Types Particularly Relevant to Earth Heritage Conservation

See Sections C.4, C.7, C.8 and C.9 of the main Handbook.

**13.** Almost any project type that may be subject to the Environmental Assessment procedures could potentially affect earth heritage interests. SNH should therefore consider potential impacts on earth heritage in all Environmental Assessment cases. However, experience indicates that particular project types frequently have significant earth heritage implications and frequently raise specific issues in the Environmental Assessment process. These are summarised in Table 3.

### Appendix 3 Table 3
### Projects Frequently Resulting in Significant Earth Heritage Impacts

| Project Type | Site Types Potentially Affected |
| --- | --- |
| Mineral extraction | Outcrops, exposures, landform, geomorphological (both static and active), river systems and stream sections; old mines and tunnels, caves and karst, unique mineral and fossil sites, mineral waste dumps, soils. |
| Landfill/landraise | Active and disused quarries, pits, cuttings, mines and tunnels, static and active geomorphological sites, caves and karst, unique mineral and fossil sites, soils. |
| Mineral restoration | Restoration of active or disused pits and quarries can affect outcrops, exposures, landform, river systems and stream sections; old mines and tunnels, caves and karst, unique mineral and fossil sites, mineral waste dumps, soils. |
| Coast protection | Coastal features, including cliffs and foreshore, and natural coastal processes including erosion and accretion. |
| Flood prevention | Coastal features, including cliffs and foreshore, and natural coastal processes including erosion and accretion, natural lochs, river systems and stream sections, soils. |
| River engineering | Riverine features, river systems, stream sections, natural lochs and soils. |
| Land drainage | Natural coastal processes, river systems, stream sections, natural lochs and soils. |
| Coastal reclamation | Coastal features, including cliffs and foreshore, and natural coastal processes including erosion and accretion. |
| Hydro schemes/reservoirs | Active and disused quarries, natural lochs, river systems and stream sections. |
| Coastal development, e.g. marinas, barrages and built developments | Coastal features, including cliffs and foreshore, and natural coastal processes including erosion and accretion. |
| Afforestation | Outcrops, exposures, landform, geomorphological (both static and active), river systems and stream sections. |
| First cultivation of uncultivated land | Static and active geomorphological sites, river systems and stream sections, soils. |
| Other land management changes | Can affect run-off, rates of erosion and accretion, sediment supplies, river systems and stream sections. |
| Dredging | Natural coastal processes including erosion and accretion. |
| Major industrial/housing or other urban developments | Outcrops, exposures, landform, geomorphological (both static and active), river systems and stream sections; old mines and tunnels, caves and karst, unique mineral and fossil sites, mineral waste dumps, soils. |

**Pressures and Impacts on Earth Science Features, Systems and Habitats**

| Pressure | Examples of on-site impacts | Examples of off-site impacts on active process systems and habitats |
|---|---|---|
| Mineral extraction (Includes pits, quarries, opencast, extraction from rivers, dunes and beaches) | Destruction of landforms and sediment records. Destruction of soils, structure and soil biodata. May have positive benefits in creating new sedimentary sections. | Contamination of watercourses. Changes in sediment supply to active process systems, leading to deposition or channel scour. Disruption of drainage network (impacts on runoff). Dust (may affect soil pH). |
| Restoration of pits quarries | Loss of exposures. Loss of natural landform. | Habitat creation. |
| Landfill | Loss of sedimentary exposures. Loss of natural landform; soil disturbance. | Detrimental effects of gases and other decomposition products on soils and soil biotas. Contamination of water courses. Contamination of groundwater. Redistribution of waste on beach/dune systems. |
| Commercial and industrial developments | Large scale damage and disruption of surface and sub-surface features including landforms and soils. | Changes to geomorphological processes downstream, arising from channelisation or water abstraction. |
| Coast protection | Loss of coastal exposures. Destruction of active and relict landforms. Disruption of natural processes. | Changes to sediment circulation and processes downdrift. |
| River management/ engineering | Loss of exposures. Destruction of active and relict landforms. Disruption of active processes. | Changes to sediment movement and processes downstream. Change in process regime. |
| Afforestation | Loss of landform and outcrop visibility. Physical damage to small scale landforms. Stabilisation of dynamic landforms (sand dunes). | Increase in sediment yield and speed of run-off from catchments during planting and harvesting. Changes to water chemistry. |
| Agriculture | Landform damage through ploughing, ground levelling and drainage. Soil compaction, loss of organic matter, reduction in biodiversity. Effects of excess fertiliser applications on soil chemistry and biodiversity. Effects of pesticides on soil biodiversity. | Changes in run-off response times arising from drainage. Episodic soil erosion leading to increased sedimentation and chemical contamination in lochs and river systems. |
| Other land management changes (e.g. drainage, dumping, construction of tracks) | Degradation of exposures and landforms. | Changes in run-off and sediment supply. Drying out of wetlands through local and distal drainage. |
| Recreation (Infrastructure, footpath develop-ment, use of all-terrain vehicles) | Physical damage to small-scale landforms and soils. Localised soil erosion. | |
| Soil pollution | Acidification of soils. Accumulation of heavy metals. | Downstream impacts on watercourses. Contamination of groundwater. |
| Soil erosion | Deterioration of landforms. | Enhanced sedimentation streams and lakes. Changes in water chemistry. |
| Climate change | Changes in active system processes. Changes in system state (reactivation or fossilisation). | Changes in flood frequency. Changes in sensitivity of landforming environments (rivers, coasts, etc.) leading to changes in types and rates of geomorphological processes (e.g. erosion, flooding). |
| Sea level rise | Changes in coastal exposures and landforms. | Changes in wider patterns of erosion and deposition. Increased flooding. |

## Assessing Significance of Earth Heritage Impacts

See Sections C.8 of the main Handbook.

**15.** Where effects on key earth heritage resources are likely to occur you should, if necessary in addition, seek the advice of your earth heritage advisors who will have experience of dealing with these issues in the Environmental Assessment process. Generally, SNH would consider earth heritage impacts to be significant where, either alone or in combination with other projects, the project would lead to:

(a) adverse or beneficial impacts on the systems or processes or features for which a geological/geomorphological SSSI had been notified;

(b) permanent or long-term change that would affect the integrity and long-term sustainable management of natural coastal processes and other natural geomorphological and hydrological systems;

(c) permanent or long-term change to the quality of the natural heritage locally or regionally as a result of the destruction or enhancement or widespread or extensive degradation or improvement of earth heritage features which have been or could merit designation as a Regionally Important Geological/Geomorphological Site (RIGS); or

(d) major constraints on or improvements to access to or interpretation of geological/geomorphological SSSI.

**16. It is particularly important that these considerations are not confined to the on-site, direct impacts of a proposal but applied equally to off-site, indirect effects such as downstream effects of river engineering works or coast protection or flood defence works or developments leading to changes in surface water run-off to natural river systems.**

# Appendix 4  Assessment of Impacts on Soils

## Background

**1.**     Soils occupy a somewhat unique position in earth heritage environmental assessment, because they are not explicitly covered by any of the existing designated area legislations in Britain. These designations are often used as the basis for assessing threats to biological, geological and geomorphological interests. A site in Wales was recently notified as a RIGS on the basis of its soils, but this is currently the only example of its kind in Britain.

**2.**     Because soils do not fit neatly into this site-based framework, they can be overlooked in environmental assessment. The position of soils at the interface between the geosphere, biosphere and hydrosphere further compounds this, as they cannot be easily compartmentalised. They also play an important part in biodiversity conservation, so it is vitally important that soils information is included as an integral part of the environmental assessment process, not only because changes to soils can have subsequent effects on other parts of ecosystems, such as vegetation composition and watercourses, but also because of the intrinsic value of the soil resource in its own right.

## Importance of Soil Considerations in Environmental Assessment
See Sections B.4 and C.3 of the main Handbook.

**3.**     Soil considerations are an essential element of the Environmental Assessment process and any significant impacts on soils should be included in an Environmental Statement.

**4.**     Annexe III of the Environmental Assessment Directive, and Schedule 4 of the EIASR 99, requires that an Environmental Statement must include a description of the aspects of the environment likely to be significantly affected by the development, including, inter alia, soil and water and the inter-relationship between them and all other aspects of the environment.

**5.**     Where significant adverse effects are identified the Environmental Statement must include a description of mitigation measures.

**6.**     Schedule 4(1) of the EIASR 99 also specifies that an Environmental Statement may include, by way of explanation or amplification, information on, inter alia:

b. the nature and quality of materials to be used in production processes;

**c. the type and quantity of expected pollutants including pollution of soils and water;**

e. the likely significant direct and indirect effects of the proposal which may result from the use of natural resources including secondary, cumulative, short, medium and long-term, permanent, temporary, positive and negative effects.

**7.** Regulations 19, 36 and 60 of the EIASR 99 provide planning authorities and the Secretary of State with the power to require the above information (and any other information in Schedule 4), having regard in particular to current knowledge and methods of assessment, where it is reasonably required to give proper consideration to the likely environmental effects of the proposed development.

**8.** Thus, soils can and should be included in an Environmental Statement and throughout the Environmental Assessment process wherever the effects of a proposal are likely to be significant. Where they are not included SNH should normally be able to request the Competent Authority to require the developer to submit the information before they grant any consent for the project. In case the Competent Authority disagrees with this approach, the precise references in the Regulations which can be used to press for soils to be included in the Environmental Statement are listed below.

| | |
|---|---|
| **Soils** | Schedule 4(3) |
| **Natural resources** | Schedule 4(4)(b) |

**9.** As it is not an offence in UK law to degrade or contaminate soil per se, the ways in which soils information is included in environmental assessments are very flexible, and can only really be influenced through various forms of guidance and advice issued by the Government and others. Examples include the Prevention of Environmental Pollution through Good Agricultural Practice code, issued by the Scottish Office, and the Forestry Commission's Forests and Soil Conservation Guidelines. Apart from the Environmental Assessment Regulations, the only other legislation that refers specifically to soil is the Sludge (Use in Agriculture) Regulations 1989, which implements EC Directive 86/278. This restricts the application of sewage sludge on agricultural land, principally on the basis of soil acidity and toxic metal concentrations in sludge and the receiving soil. Planning legislation provides little additional support for soils, as it deals principally with land as space, and not the soil functions listed below.

## Soil Functions

**10.** For assessment purposes, soils can be considered to have six general functions:

- production of biomass
- filtering, storage and transformation of substances
- support of biodiversity
- provision of a physical base–for plants, buildings and infrastructure
- provision of raw materials
- protection of heritage (i.e. archaeological) sites.

These functions can be translated into either economic or ecological forms of land use.

## Soil Heterogeneity

**11.** Different soil types have their own characteristic properties, which affect the significance and magnitude of impacts. Some soils are relatively robust and are able to support a wide range of potential applications, whereas others can only be utilised in more limited ways. Within any given area, there is likely to occur a variety of soils, which can pose planning problems, often leading to some soils being exploited in ways for which their properties are unsuited. Further complexities are introduced by the fact that, unlike geological exposures or landforms, which occupy distinct areas of the landscape and are generally fairly easy to assess, soils form a continuous pattern over the land surface and are for the most part hidden from view. All of these factors combine to create very specific requirements for environmental assessment of soils.

## Source of Information
See Sections C.4, C.5 and C.6 of the main Handbook.

**12.**   In order for informed decisions to be made, an adequate source of data is a necessity. On a national scale, Scotland is well covered by soil maps produced by the former Soil Survey of Scotland (now part of the Macaulay Land Use Research Institute (MLURI) in Aberdeen), with complete coverage at 1:250,000 (countrywide), 1:63,360 (lowland areas) and 1:50,000 (upland areas). In addition, MLURI holds a comprehensive database of over 12,000 soil profile descriptions, collected concurrently with the mapping programme.

**13.**   This data is of sufficient detail for assessment of land with reference to broad categories of land use. On a more local scale, though, existing spatial soil data tend to be patchy, of variable quality through being obtained by a range of methods, and difficult to access, often being unpublished and held by a number of different organisations and individuals. There is a particular scarcity of data in urban and peri-urban areas, as soil surveys have traditionally been carried out almost solely for agricultural purposes. As most environmental assessments are made at the more site specific level, it is essential that the authorities involved seek appropriate advice where it is evident that soil factors will be integral to the assessment. The scoping stage is of particular importance here, as the opportunity to raise the issue of effects on soils at an early stage.

## References

**14.**   Useful sources of soil information for environmental assessment include:
*Forestry Commission Forests and Soil Conservation Guidelines*, London, HMSO.
Scottish Office (1992) *Prevention of Environmental Pollution from Agricultural Activity: Code of Good Practice*. Edinburgh, Scottish Office
(a revised and updated version of the Code to be published).
*Soil Survey of Scotland* (1982) 1:250,000 Soil Survey of Scotland maps and handbooks 1–7. Aberdeen, The Macaulay Institute for Soil Research.
*Soil Survey of Scotland (1984) Organisation and Methods of the 1:250,000 Soil Survey of Scotland*. Aberdeen, The Macaulay Institute for Soil Research.
Predicting Soil Impacts: Projects Likely to Give Rise to Impacts on Soils,
see Sections C.4, C.7 and C.8 of the main Handbook.

**15.**   Some of the main project types likely to give rise to impacts on soils in environmental assessment (see Figure 1 below) and which can be directly relevant to the functional capacity, sensitivity, vulnerability and general condition of soils include:

- location of developments (e.g. sewage works, hazardous installations, landfill sites)
- other industrial developments
- urban and infrastructure development

- reclamation of contaminated and derelict land
- land instability
- land drainage
- mineral extraction
- archaeological excavations
- land restoration
- recreation (e.g. footpaths, sports facilities)
- land use changes associated with forestry
- land use changes associated with agriculture.

## Predicting Soil Impacts: Impacts on Soils

See Sections C.4, C.7 and C.8 of the main Handbook.

**16.** The impacts of these projects on soil properties and processes can include:

- erosion
- pollution, e.g. from heavy metals, organic compounds, industrial wastes, fertilisers, pesticides
- changes in pH
- loss of or reduction in biodiversity
- loss of organic matter
- compaction
- structural deterioration
- homogenisation and loss of characteristic horizons
- physical and chemical changes associated with topsoil stripping and storage
- changes associated with land restoration
- decline in fertility
- destruction or modification of palaeosols
- changes to soil water regime
- removal or alteration of parent material
- loss or burial of soil.

**17.** Figure 1 below summarises the main pressures on soils and examples of the various types of on-site and off-site impacts they may cause.

## Soil Properties: Mitigating Measures

See Section C.9 of the main Handbook.

**18.** By matching as far as possible particular developments with appropriate soils, the consequences of many of these impacts can be minimised. In this context, environmental assessment involves the consideration of key soil properties and characteristics in relation to the proposed development or change of land use. Some of the more important soil properties that should be considered in mitigation measures are:

- texture
- structure
- organic matter content
- pH
- nutrient status
- depth—both total and of individual horizons
- parent material characteristics
- horizontation (i.e. nature and arrangement of individual horizons)
- salinity
- stoniness
- soil water regime.

**Examples of Pressures and their Impacts on Soils**

| Pressure | On-site impacts | Off-site impacts |
|---|---|---|
| Reclamation of contaminated land | Disposal of contaminants. Changes in chemistry. Lack of suitable quality soil. | Leakage of contaminants to watercourses. |
| Location of developments | Soil loss. Contamination. Structural damage. Changes to soil water regime. Disposal of wastes. Effects on soil biota. | Leakage of contaminants to watercourses. Groundwater contamination. Effects of waste products on vegetation. |
| Urban and infrastructure development | Soil loss or burial. Contamination. Structural damage. | Ground and surface water contamination. |
| Land instability | Shrinkage/swelling of clays. Compaction. Erosion. | Movement of soil off-site. |
| Land drainage | Oxidation of organic matter. Physical damage. Soil water changes. Effects on pH. | Sedimentation of water courses. Changes to water chemistry. |
| Mineral extraction | Loss of soil. Physical damage. Effects on biota. Contamination. Soil stripping and storage. | Contamination of water courses. Changes to sediment load. |
| Archaeological excavations | Damage to palaeosols. | |
| Land restoration | Problems associated with reinstatement of previous soil conditions. | Changes to water chemistry. |
| Recreation | Erosion. Compaction. Loss of organic matter. | |
| Forestry | Erosion. Changes to pH. Changes to horizons. Changes to soil water. Effects on soil biota. | Increased sediment yield. Changes to run-off. Changes in water chemistry. |
| Agriculture | Loss of organic matter. Erosion. Changes to nutrient status. Compaction. Structural damage. Effects on biodiversity. pH changes. Homogenisation. | Pollution of groundwater. Pollution of surface water. Increased sediment yield. |

# Appendix 5  Outdoor Access Impact Assessment

## Introduction

There is no precise definition of 'outdoor access'. Rather, it is a diverse collection of activities which are linked by common values and by a dependence on open air settings for their practice or enjoyment. It can range from walking to windsurfing to bird watching. Outdoor access can be undertaken for a variety of purposes including recreation, educdation, socialising, health benefits and travel from one place to another. The types of recreation in which SNH has a particular interest are as follows.

● recreation that is dependent on, or draws inspiration from the enjoyment of the qualities of the outdoors;

● recreation that is practised informally and non-competitively;

● recreation that is accessible to and practised by the general public, without the need for membership of groups or societies in order to practice that activity.

Whatever the activity, SNH takes an interest in all types of recreation which take place out of doors, especially where they make use of natural resources or have effects on them, or on other people's enjoyment of their own recreation.

**1.**     This Appendix explains in more detail the issues likely to arise in the Environmental Assessment process in respect of outdoor access. There are no published Environmental Assessment techniques or good practice methods relating specifically to outdoor access impact assessment. Outdoor access issues are often overlooked or understated in published Environmental Statements unless a particularly important facility is involved. This Appendix is to enable SNH to remedy such deficiencies.

**2.**     It should be borne in mind that recreational developments may themselves create impacts on the natural heritage. These impacts will need to be assessed by the general procedures set out in this guide, and include adverse effects on, or opportunities for access to the recreation being practised on land to be developed or adjacent to it.

**3.**     The general procedures of assessment described in the main text of this Handbook are relevant and applicable to outdoor access issues. Equally, outdoor access issues should be an integral consideration at every step in the process. This Appendix:

a.     sets out the importance of outdoor access considerations in Environmental Assessment;

b.     summarises the main outdoor access provisions relevant to the Environmental Assessment process;

c.     identifies the main or typical threats to outdoor access, i.e. the main potential impacts, and project types particularly relevant to outdoor access provision; and

(d)     provides general advice on assessing the significance of outdoor access impacts.

**4.**     There are very close relationships between the likely effects of development on visual amenity, guidance on which is found in Appendix 1 of this Handbook, and the extent to which people's enjoyment will be impaired, either in the generality or, in many cases, when engaging in open air recreation. This Appendix, however, deals with issues that arise when developments impinge on the ability of people to engage in open air recreation or on the facilities used by them rather than what may be seen of the development from the place at which people are taking their leisure.

## Enjoyment of the Natural Heritage

**5.**     SNH's responsibilities for enjoying the natural heritage are founded in the 1967 Countryside (Scotland) Act, and in its enabling legislation, the 1991 Natural Heritage (Scotland) Act. In the legislation, the word enjoyment is primarily about the use of the countryside for open air recreation, with the 1967 Act providing the local authorities (and also SNH) with a range of powers and duties to facilitate better access and the provision of facilities. The National Parks (Scotland) Act 2000 gave to national park authorities powers similar to those of local authorities. Part 1 of the Land Reform (Scotland) Act 2003 established statutory access rights to most land and inland water, subject to these rights being exercised responsibly, and also introduced very specific duties and powers for local authorities and national park authorities for upholding access rights and for planning and managing access. This Act has been effective since 9th February 2005. People can also enjoy the countryside as part of everyday travelling to work or for social reasons; people enjoy both extensive and small scale elements of the natural heritage and it is also possible to enjoy the values of the countryside at a distance, as an important existence value, through which it is sufficient for people to know that valued places exist and are unaffected by adverse changes.

**6.**     In this way, enjoyment can encompass values which underlie both the physical aspects of recreation and the varied aesthetic pleasures that people find in the outdoors. SNH's prime role is with the informal pursuits, but we should also take a positive stance on behalf of the active and organised pursuits which primarily fall under Sport Scotland's remit, always assuming that these activities themselves are being practised in ways which do not cause adverse effects on natural resources. Some recreational activities do fall outside SNH's remit. These include field sports such as angling and shooting which are normally a form of private or commercial use of land or water, and fall outside issues of access rights. Also the way in which these recreational activities are practised, and their commercial links and special management needs put them well beyond SNH's remit to facilitate public enjoyment.

**7.**     In assessing the effects of a development on access, a distinction should be made between the access itself, which is the ability to make use of a site or route, and accessibility, which is the ease with which access can be taken. In different settings, these factors may have different levels of significance. In settings close to where people live we are usually concerned to enhance both access and accessibility but in a remoter setting in open country, access may not be a major factor and greater accessibility of less certain value. Assessment should always consider the needs of recreation dispersed in the countryside as well as at facilities, and the importance of local networks (and threats to the loss of key linkages) always borne in mind.

## Importance of Outdoor Access Considerations in Environmental Assessment

**8.**     Outdoor access considerations are an essential element of the Environmental Assessment process and any significant impacts on outdoor access features and sites must be included in an Environmental Statement.

**9.**     Annexe III of the Environmental Assessment Directive, and Schedule 3 of the EASR 1988, require that an Environmental Statement must include 'a description of the likely significant effects, direct and indirect, on the environment of the proposed development, explained by reference to,' inter alia,

a.     human beings
h.     the landscape and
i.     interactions of these with each other and with wildlife, the air, soils, and the climate

and

j.     material assets and
k.     the cultural heritage.

**10.**     Where significant adverse effects are identified, the Environmental Statement must include a description of mitigation measures (see para 22).

**11.**     Schedule 3(3) of the EASR 88 also specifies that an Environmental Statement may include, by way of explanation or amplification, information on, inter alia,

c.     the type and quantity of expected pollutants including noise, vibration, light, heat, and radiation;
e.     the likely significant direct and indirect effects of the proposal which may result from the emission of pollutants and the creation of nuisances, including secondary, cumulative, short, medium and long-term, permanent, temporary, positive and negative effects.

**12.**     Regulations 22, 43, 54 and 68 provide planning authorities and the Scottish Ministers with the power to require the above information (and any other information in Schedule 3(3)), having regard in particular to current knowledge and methods of assessment, where it is reasonably required to give proper consideration to the likely environmental effects of the proposed development.

**13.**     Thus, outdoor access interests can and should be included in an Environmental Statement as a crucial element of the interaction of human beings with the environment often involving resources of a physical or cultural nature. They should be reflected throughout the Environmental Assessment process, wherever the effects of a proposal are likely to be significant. Where they are not included SNH should normally be able to request the Competent Authority to require the developer to submit the information before they grant any consent for the project. In case the Competent Authority disagrees with this approach, the precise references in the Regulations which can be used to press for outdoor access matters to be included in the Environmental Statement are listed below.

## Appendix 5 Table 1   References in Regulations

| Impacts | Topic in Regulations | Reference in Regulations |
|---|---|---|
| Effects on people | Human beings | Sch. 3(2)(c)A |
| Effects on the Landscape | Landscape | Sch. 3(2)(c)H |
| Effects of pollutants on facilities | Water<br>Air<br>Soil<br>Pollutants, residues, emissions | Sch. 3(2)(c)E<br>Sch. 3 (2)(c)F<br>Sch. 3(2)(c)D<br>Sch. 3(3)(c) |
| Effects of noise, vibration etc. on facilities | Pollutants, residues, emissions | Sch. 3(3)(c) |
| Creation of nuisance | Pollutants and nuisances | Sch. 3(3)(e) |

## Outdoor Access Facilities Relevant to Environmental Assessment

See Sections D.4, D.7 and D.8 of the main Handbook.

**14.**   The potential effects of a project on outdoor access interests will usually depend on the following main considerations:

a.      the type of outdoor access facility;

b.      the type of project, including its nature, scale, location, duration etc.; and

c.      the nature of recreation practised at the site or facility.

**15.**   Impact assessment therefore needs to take account of the differing issues and objectives for outdoor access facilities. Table 2 below summarises the types of countryside access facilities relevant to the Environmental Assessment process.

## Appendix 5 Table 2

Outdoor Access Resources Relevant to Environmental Assessment

| Area based facilities | National Park<br>Regional Park<br>Country Park<br>Picnic Sites and other roadside facilities<br>(now repealed)<br>Areas subject to S.49A Management Agreements including public access<br>National Nature Reserve<br>Local Nature Reserves<br>Local open space and green space<br>Inland lochs and reservoirs |
|---|---|
| **Linear access facilities** | Core Paths and the wider paths network available through access rights<br>Long Distance Routes, regional routes, National Cycle Network<br>Public rights of way<br>Path Agreements (S.30 of the Countryside (Scotland) Act and S.21 of the Land Reform (Scotland) Act<br>Permissive paths and routes on land where access rights do not apply |

**16.** The effects of any major development on people's ability to enjoy open recreation in the countryside can arise in a number of different ways, as shown in Table 3 below.

## Appendix 5 Table 3
**Effects of Development on People's Ability to Enjoy Open Recreation in the Countryside**

| Type of effect | Implications |
|---|---|
| Effects on the intrinsic quality of the resources enjoyed by people. | Aesthetic changes, mainly visual and aural as considered in earlier sections of this Appendix. However, there are recreational values in solitude, challenge and hazard, enjoyment of wildlife and habitats or in the sociability enjoyed in the more gregarious pursuits, which can be affected by development. |
| Direct effects on the facilities or infrastructure used to take access or for the practice of recreation. | Restriction of access to facilities, barriers, physical restrictions or limitations on the use of the site or facility, or even its loss. |
| Effects on the practice of recreations. | Restrictions or limitations on the kinds of recreations pursued, or in the ways in which they are practised, and limitations on specific recreations, say by reduction in available space. |
| Foreclosure on options for future access development. | Any of the foregoing which might affect proposals either formalised and recorded statutory plans or local access and recreation strategies or less formerly known about, which might limit development of future options in enhancing the supply or quality of recreation opportunity for a community. |
| Implications for public safety. | These might arise from the development itself from the relocation of facilities to a less suitable location, from the intensification of use, or from the mixing of recreations previously having more space for their own use. |
| Restrictions on the less able. | Reductions to accessibility may lead to effects on the enjoyment of the disabled, the elderly or people who are otherwise disadvantaged (say, those without access to private motor transport). |
| Effects on particular recreations. | These will be assessed according to the specific circumstances, but might include issues such as a loss of access to boat launching, restrictions to a bridleway network, on the loss of key link routes in an access network etc. |

## Types of Impact
See Sections D.4, D.7 and D.8 of the main Handbook

**17.** All significant potential effects on outdoor access facilities should be assessed. Generally, effects, or impacts, are likely to fall into one or more of the types summarised in Table 4 below. For each type, an example of potential impacts is given.

## Appendix 5 Table 4
**Examples of Potential Outdoor Access Impacts**

| Type | Example | Timescale | Reversibility | Comments |
|---|---|---|---|---|
| Loss/closure/ extinguishment | Landtake from route or facility for built development | Permanent | Usually irreversible | Usually adverse can be cumulative |
| | Permanent closure of right of way at motorway | Permanent | Irreversible | |
| | Temporary closure for mineral extraction | Short to medium-term | Reversible | |
| Diversion | Hydro scheme or trunk road requires diversion of path | Permanent | Irreversible | Usually adverse can be cumulative |
| | Mineral extraction requires diversion of path | Long-term | Reversible | |
| | Waste disposal requires diversion of path | Medium-term | Reversible | |
| | Building construction works require diversion of path | Short-term | Reversible | |
| Reduction in amenity | Industrial plant/factory causes noise or smell to section of Long Distance route | Permanent | Irreversible | Usually adverse can be cumulative |
| | Mineral working causes noise, dust or vibration to country park | Long-term | Reversible | Usually adverse can be cumulative |
| Enhancement of amenity | Golf course adjacent to a country park reclaims derelict land | Permanent | Irreversible | Usually beneficial |
| Intrusion | Telecommunications mast in wildland area | Medium to long-term | May be reversible | Usually adverse |
| Obstructing access routes | Closure of paths to, e.g., viewpoints and natural features | Various timescales | Usually reversible | Usually adverse |
| Enhancing access | Provision of access and/or interpretation | Various timescales | Usually reversible | Usually beneficial |
| Changes to setting and context | Built development adjacent to Regional Park | Permanent | Irreversible | Usually adverse can be cumulative |

## Project Types Particularly Relevant to Countryside Access Conservation

See Sections D.4, D.7 and D.8 of the main Handbook.

**18.** Almost any project type that may be subject to the Environmental Assessment procedures could potentially affect outdoor access interests. SNH should, therefore, be alert to the potential impacts on outdoor access in all Environmental Assessment cases. However, experience indicates that particular project types frequently have significant outdoor access implications and frequently raise specific issues in the Environmental Assessment process. These are summarised in Table 5.

### Appendix 5 Table 5
**Projects Frequently Resulting in Significant Outdoor Access Impacts**

| Project Type | Facilities Potentially Affected |
| --- | --- |
| Mineral extraction | Adverse effects on adjacent facilities from noise, dust, vibration and visual impact and can require closure or diversion of linear facilities for long periods of time. |
| Landfill/land-raise | Adverse effects on adjacent facilities from noise, dust, smell and visual impact and can require closure or diversion of linear facilities for medium periods of time. |
| Flood prevention | Direct and indirect impacts on amenity of area facilities and can lead to permanent closure or diversion of linear routes. |
| Windfarms | Can change perception and amenity of both area and linear facilities through visual and noise impacts, access tracks can interfere with/or facilitate public access, general deterrent/attractor effects. |
| Hydro schemes/reservoirs | Can affect wildland qualities of remote areas, interfere with pre-existing access and adversely affect amenity of routes along rivers/lochsides. Also can affect the ability to undertake, and the quality of, recreation on or in rivers. |
| Powerlines, masts and other pylons | Effects on valued landscapes especially in remote countryside. |
| Afforestation | Can affect wildland qualities of remote areas, interfere with pre-existing access and adversely affect amenity of area facilities and routes in the uplands. |
| Roads | Major road proposals often sever access routes and may lead to extinguishment or diversion. Area facilities can be adversely affected by noise and visual intrusions. |
| Major urban developments | Can reduce amenity value of area facilities and increase pressures for use on vulnerable areas. Can lead to extinguishment or diversion of rights of way. |
| Rural industrial and statutory undertaker developments | Effects on valued landscapes, loss of rural character, inappropriateness in the setting, loss or division of linear access routes, pollution. |
| Recreational developments | Implications for other recreational users/the site or area resource, affecting the quality of enjoyment by others, affecting levels of use of an area or accessibility to it. |

# Assessing Significance of Outdoor Access Impacts

See Sections C.3, D.7, D.8 and E.10 of the main Handbook (duplicated below).

**19.** Reference is made to sections C.3, D.7 and D.8 of the main text of this Handbook.

**20.** Where effects on key outdoor access resources are likely to occur you should seek the advice of your outdoor access advisors who will have experience of dealing with these issues in the Environmental Assessment process. Generally, SNH would consider outdoor access impacts to be significant where, either alone or in combination with other projects already approved, the project would lead to:

a. permanent or long-term effects on the resources on which enjoyment of the natural heritage depends, in particular where facilities have been provided by SNH or others under statutory powers;

b. permanent or long-term change that would affect the integrity and long-term sustainable management of facilities which were provided by SNH or others under statutory powers;

c. where there are recreational resources for open air recreation pursuits affected by the proposal which have more than local use or importance, especially if that importance is national in significance;

d. major constraints on or improvements for access or accessibility to designated natural heritage sites;

e. where mitigation and/or compensatory or alternative recreational provision is considered to be inadequate.

**21.** Assessment of the scale of prospective effects from a development on recreation and access can be considered against a range of criteria, not all of which will apply in any one circumstance. The main factors to be borne in mind are as follows.

## Appendix 5 Table 6
### Assessment of the Scale of Effects of Development on Outdoor Recreation and Access

| Assessment Factor | Commentary |
| --- | --- |
| Obtrusiveness or the scale of the effect in the context in which recreation takes place | Thus noise effect might be accommodated or open to mitigation in an urban edge setting, but much less acceptable or even beyond amelioration in remoter countryside. |
| The intensity, frequency of occurrence, timing, or performance of the effect | These will be important factors in assessment of the acceptability of effects on recreation. At the less intense levels of effect, the outcomes may be acceptable or open to mitigation either in intensity or through time limitations on certain activities within the development. |
| Potential for the effects to augment and scale and frequency of occurrence | This is a precautionary point of reasonable anticipation of how effects might increase in scale over the years and thereby make mitigation ineffective. |
| The extent of local supply of access and recreation opportunities | This involves an assessment of the extent to which a community has a shortage (or ample supply) of facilities which may be made worse by a development, allowing for the potential for compensatory provision through the creation of alternative opportunities. |

| Assessment Factor | Commentary |
|---|---|
| Scarcity value of the recreation resource | This factor recognises there are considerable geographic imbalances in the supply of recreation opportunities, and where a resource is in short supply then less compromise may be feasible. As examples, some parts of the country are very poorly endowed with accessible open water space. |
| Recognition of the recreation opportunities spectrum (ROS) | The recreation opportunities spectrum is a basic principle of recreation planning that provision should be made for people's recreation needs along a range which provides for gregarious, active and some noisy recreations at one end of the scale, to opportunities for the enjoyment of solitude and quiet enjoyment at low densities of participation. This ensures high quality provision for small numbers of people at one end of the spectrum as important as providing for the many at the other end. |

## Opportunities for Mitigation

See Sections D.9, E.7 and F.3 of the main Handbook.

**22.**    Developments need not just create effects on recreation resources, on the practice of recreations, or on the potential for the future needs of a community. Opportunities may arise from new developments, either directly as a consequence of new access being created or through the opportunity to manage or plan for better recreation provision. In some cases recreation itself may be having effects on natural resources which are not managed effectively, or too sensitive to bear increased use, in cases where accessibility is enhanced. The nature of mitigation will have to be tailored to resolving or compensating for the predicted adverse effects described earlier, but some general issues to consider are summarised in Table 7 below.

### Appendix 5 Table 7
**Mitigation Measures for Outdoor Access and Recreation**

| Measure | Commentary |
|---|---|
| Realignment of access | This can be a simple and very acceptable measure, provided that major diversion of use is not proposed. Excessive diversion may lead to non-compliance or be inadequate to serve certain interests, especially the disadvantaged and disabled, or simply may be too distant for visitors' real needs. |
| Compensatory provision | Best acceptance of alternative provision will arise where the overall benefits are perceived by the visiting public to outweigh the losses, particularly if both access and accessibility are enhanced. |
| Reduction of the intensity, frequency or timing of the effects | Mitigation of this kind is always potentially acceptable, provided that the amelioration is realistic, can be guaranteed and is not in due course overtaken by a greater intensity of use at the development site, and provided that the adverse effects are not wholly incompatible with the nature of recreational use and its setting, in which case, limitations will probably not be effective. |
| Enhanced management provision for the recreational use of the area or site | Likely to provide beneficial mitigation, especially where recreation management was absent or weakly provided for, and perhaps even a cause of adverse effects itself on natural resources. However, new management regimes in any area where recreation is a significant use of land should be the subject of consultation with recreational interests, especially where the development itself is a matter of controversy as a result of its projected effects on the enjoyment of the natural heritage. |

# Appendix 6   Effects on the Marine Environment

## Introduction

**1.** This Appendix explains in more detail the issues likely to arise in the EIA process in respect of the marine environment. Compared to terrestrial interests, there are very few published EIA techniques or good practice methods relating specifically to marine impact assessment (see list of references below). Marine issues are often overlooked in published Environmental Statements and, unless a European Marine Site is involved, Competent Authorities may also overlook potential impacts on the marine environment. Consequently, consideration of these impacts may be absent or inadequate at any stage in the EIA process and one of the key objectives of the guidance in this Appendix is to enable SNH to remedy such deficiencies.

**2.** The EIA process described in the main text of this Handbook is entirely relevant and applicable to impact assessment on the marine environment. Equally, marine issues should be an integral consideration at every step in the process. This Appendix:

  a. sets out the importance of marine considerations in EIA;

  b. summarises the designations relating to marine areas and their conservation objectives relevant to the EIA process;

  c. identifies the main or typical threats to the marine environment, ie. the main potential impacts, and project types particularly relevant to the marine environment; and

  d. provides general advice on assessing the significance of impacts on the marine environment.

## References

**3.** Reference is made here to the following publications:

A. Crown Estate Commissioners, Feb 2000, *Environmental Assessment Guidance Manual for Marine Salmon Farmers*, CEC

B. SNH, March 2000, *Marine Aquaculture and the Landscape: The siting and design of marine aquaculture developments in the landscape*, SNH

C. Marine Biological Association of the UK, *Marine Life Information Network (MarLIN)*, ongoing website at http://www.marlin.ac.uk

D. Campbell, J.A., 1993, *Guidelines for assessing marine aggregate extraction.* MAFF Laboratory Leaflet Number 73, Directorate of Fisheries Research, Lowestoft, 1993 ISSN 0143 8018

E. Vella, G. *et. al.* (University of Liverpool, Centre for Marine and Coastal Studies

Environmental Research and Consultancy) 2001, *Assessment of the effects of noise and vibration from offshore windfarms on marine life*, ETSU W/13/00566/REP. DTI publication URN 01/1341

## Importance of Marine Considerations in EIA

See also text in the main Handbook.

**4.**    Marine considerations are an essential element of the EIA process and any significant impacts on marine features and sites must be included in an Environmental Statement.

**5.**    Annexe III of the EIA Directive, and Schedule 4 Part I of the EIASR 99, require that an Environmental Statement must include

*a description of the aspects of the environment likely to be significantly affected by the development including in particular ... fauna, flora, ... water, air, climatic factors, ... landscape ... and the interrelationship between the above factors ...*

These clearly are intended to include the marine environment.

**6.**    Where significant adverse effects are identified the Environmental Statement must include a description of mitigation measures.

**7.**    Schedule 4(4) of the EIASR 99 also specifies that an Environmental Statement may include:

*A description of the likely significant effects of the development on the environment, which should cover the direct effects and any indirect, secondary, cumulative, short, medium and long term, permanent and temporary, positive and negative effects of the development resulting from:*

a   *the existence of the development;*
b   *the use of natural resources;*
c   *the emission of pollutants, the creation of nuisances and the elimination of waste.*

**8.**    The Regulations provide Competent Authorities and the Scottish Ministers with the power to require the above information (and any other information in Schedule 4(4)), having regard in particular to current knowledge and methods of assessment, where it is reasonably required to give proper consideration to the likely environmental effects of the proposed development.

**9.**    Thus, all marine interests can and should be included in an Environmental Statement and throughout the EIA process wherever the effects of a proposal are likely to be significant. Where they are not included SNH should normally be able to request the Competent Authority to require the developer to submit the information before they grant any consent for the project.

## Marine Site Designations Relevant to EIA

**10.**    The potential effects of a project on marine interests will usually depend on 2 main considerations:
    a. the type of marine interest; and

b. the type of project, including its nature, scale, location, duration etc.

**11.** Impact assessment therefore needs to take account of the differing issues and conservation objectives for marine sites. Table 1, on the next page, summarises the natural heritage designations relevant to marine areas.

## Appendix 7 Table 1
### Marine Site Designations Relevant to EIA

| Designation | Interest/Purpose of Designation |
|---|---|
| Marine Nature Reserve (MNR) | Conservation and study of marine flora or fauna or geological or physiographical features. |
| European Marine Site (Marine SPAs and Marine SACs) | Comprises those parts of Special Protection Areas (SPAs) under the EC Birds Directive and Special Areas of Conservation (SACs) under the EC Habitats Directive that lie below Highest Astronomic Tide (HAT). The interests for which the sites are designated are, in the case of SPAs, the bird species listed in Annexe 1 of the Birds Directive and, in the case of SACs, the habitats and species listed in Annexes I and II of the Habitats Directive respectively (and also giving added protection to species listed in Annexes IV and V) that are specified in the citation for the classification/designation. European Marine Sites are subject to the protection and management provisions of the Habitats Regulations 1994. |
| Sites of Special Scientific Interest (SSSI) with marine features | The protection and management of sites which, in the opinion of SNH, are of special scientific interest by reason of their flora or fauna or geological or physiographical features. SSSI will normally extend down to Mean Low Water Mark Ordinary Spring Tides (MLWMOST) but the interests may extend beyond this level down to Lowest Astronomic Tide (LAT) or to sub-tidal areas. |
| Marine Consultation Areas | A non-statutory designation intended to assist in the protection and conservation of marine areas which are of high conservation value but not otherwise designated |
| National Scenic Areas | Designated by the Scottish Ministers to conserve the natural beauty and amenity of some of the finest landscapes in Scotland, several of which include coastal areas and some include extensive areas of sea as well as land. |

## Types of Impact
See also relevant sections in the main Handbook and Appendix 4.

**12.** All likely significant effects (or impacts) on marine interests should be assessed. Projects could affect the ecology, biology, geology, geomorphology, visual (both landscapes and seascapes), cultural and amenity value or accessibility of the marine environment and SNH could be concerned about any of these. For example, projects can have:

Direct effects: such as
  landtake with consequent loss of habitat from intertidal or subtidal areas;
  severance or fragmentation of areas e.g. by the construction of barriers or causeways;
  burial of marine flora and fauna by deposits on the sea bed;
  loss of marine flora or fauna and disturbance to habitats caused by extraction

of material from the sea bed;

visual intrusion caused by conspicuous and uncharacteristic structures detracting from visual amenity;

loss of small jetties or quays which facilitate quiet recreational enjoyment of coastal areas; or

Indirect effects: such as

changes in tidal prisms or sediment budgets in natural systems caused by a one-off 'capital' dredge or frequent maintenance dredging;

interruption or other changes to natural coastal processes, e.g. by coast protection works;

noise disturbance to birds, e.g. from land based industrial activity or from increased use of powered boats;

vibration disturbance to fish and marine mammals, e.g. from blasting or drilling operations;

changes in sediment erosion or deposition caused by increased navigation;

increased disturbance to marine flora and fauna caused by increased levels of recreational diving/sub-aqua activities.

**13.** In the marine environment it is particularly important to consider cumulative impacts. One discharge to the sea or one small physical change to coastal processes may be acceptable on its own, but in combination with the effects of other projects could cause a significant adverse effect.

**14.** The relative paucity of information about some aspects of the marine environment may also mean that the precautionary principle may need to be invoked more often in marine EIA cases (see section F.1 of the main text of this Handbook).

## Project Types Particularly Relevant to Marine Conservation

See also relevant sections of the main Handbook.

**15.** Almost any project type that may be subject to the EIA procedures could potentially affect marine interests. SNH should therefore consider potential impacts on marine conservation in all EIA cases. However, experience indicates that particular project types frequently have significant marine implications and frequently raise specific issues in the EIA process. These are summarised in Table 2. It should also be appreciated that the different life stages of a project may have different effects on the marine resource; these different life stages are described in the main Handbook.

## Projects Frequently Resulting in Significant Marine Impacts

| Project Type | Marine Natural Heritage Interests Most Likely to be Potentially Affected* |
|---|---|
| Marine dredging for mineral extraction (construction and maintenance) <br><br> Marine dredging for navigation | Water quality, flora, fauna, marine habitats, geological features and natural coastal and maritime systems and processes including sediment drift, erosion and accretion |
| Deposit of dredgings at sea or on the coast | As above plus potential landscape and visual amenity and access to the coast |
| Waste management and disposal of waste at sea | Water quality, flora and fauna, marine habitats, natural coastal and maritime systems and processes |
| Aqueous and other liquid discharges to the sea including waste water treatment work discharges | Water quality (including temperature), flora and fauna, marine habitats, natural coastal and maritime systems and processes, visual amenity |
| Gaseous emissions likely to be deposited on the marine environment | Water quality, flora and fauna, marine habitats |
| Radioactive discharges into the sea | Water quality, flora and fauna, marine habitats |
| Laying cables, pipes etc. on or in the sea bed | Flora, fauna, marine habitats, geological features and natural coastal and maritime systems and processes including sediment drift, erosion and accretion |
| Land claim from the sea <br><br> Coast Protection Schemes <br><br> Flood Prevention Schemes | Water quality, flora, fauna, marine habitats, geological features and natural coastal and maritime systems and processes including sediment drift, erosion and accretion; landscape and visual amenity and access to the coast |
| Transport infrastructure including bridges and causeways | Flora, fauna, marine habitats, geological features and natural coastal and maritime systems and processes including sediment drift, erosion and accretion; landscape and visual amenity and access to the coast |
| Coastal Development, e.g. marinas and built developments | Water quality, flora, fauna, marine habitats, geological features and natural coastal and maritime systems and processes including sediment drift, erosion and accretion; landscape and visual amenity and access to the coast |
| Energy projects including marine wind turbine generators, wave energy generators and tidal barrages | Water quality, flora, fauna, marine habitats, geological features and natural coastal and maritime systems and processes including sediment drift, erosion and accretion; landscape and visual amenity and access to the coast |
| Marine aquaculture | Water quality, flora, fauna, marine habitats, natural maritime systems and processes; landscape and visual amenity and access to the coast |

*N.B. The interests listed here are intended to be illustrative of the range and nature of natural heritage interests potentially affected. The Table should not be used as a 'checklist' and all projects should be carefully scoped for all potentially significant effects on the natural heritage and wider environment where appropriate.

## Assessing Significance of Marine Impacts

**16.**    Where effects on key marine resources are likely to occur you should, if necessary in addition, seek the advice of your marine advisors who will have experience of dealing with these issues in the EIA process. Generally, SNH would consider marine impacts to be significant where, either alone or in combination with other projects, the project would lead to:

a. adverse or beneficial impacts on the systems or processes or features for which a site had been notified or designated;

b. permanent or long-term change that would affect the integrity and long-term sustainable management of natural coastal processes and other natural marine systems;

c. permanent or long-term change to the quality of the natural heritage locally or regionally as a result of the destruction or enhancement or widespread or extensive degradation or improvement of marine habitats, species populations or features.

**17.   It is particularly important that these considerations are not confined to the on-site, direct impacts of a proposal but applied equally to off-site, indirect effects such as outfalls or coast protection or flood defence works or developments leading to changes in surface water run-off to rivers/estuaries etc.**

**SCOTTISH EXECUTIVE**

Victoria Quay
Edinburgh EH6 6QQ

Development Department
Planning Division

Telephone: 0131-244 7710
Fax: 0131-244 7083
John.Gunstone@scotland.gsi.gov.uk
http://www.scotland.gov.uk

Heads of Planning

CC. CoSLA
      SSDP

Our ref:          PGD/5/12

June 2002

_____  _____

Dear Sir/ Madam

## ENVIRONMENTAL IMPACT ASSESSMENT (EIA) DIRECTIVE
**1) Minimum Requirements of the Regulations**
**2) Outline Planning Applications**

Increasingly over the past few years planning decisions have been challenged on grounds that the planning authority has not, or has not properly, complied with the requirements of the European Community Directive on the assessment of the effects of certain public and private projects on the environment (the EIA Directive). Local authorities need to ensure compliance with the Directive so that environmental impacts can be properly considered. For all concerned, challenges are costly and time consuming. They delay and frustrate the planning system and do little to encourage belief in its efficiency.

We cannot prevent such challenges. But careful application of the Regulations that implement the Directive will help minimise the number of challenges and should also limit the likelihood of a successful challenge.

### Minimum Requirements of the Regulations

Attached to this letter is a note in the form of a Q&A brief that sets out minimum requirements of the Environmental Impact Assessment (Scotland) Regulations 1999 (SSI 1999/1) (the EIA Regulations) in so far as they relate to planning. It also highlights some EIA related issues that have arisen in recent Court cases and indicates actions that your planning staff can take to avoid similar difficulties. The Court cases are English cases, but given the similarities in the EIA regimes north and south of the border, they are relevant to the operation of the EIA Regulations in Scotland. I would be grateful if you could disseminate this letter and the accompanying note widely to planning officers within your Authority.

### Outline Planning Applications

The attached Q&A contains advice on handling outline planning applications where EIA is required. It points to how an outline planning permission should be constructed to ensure the development is within the parameters of the environmental information provided via the EIA process. However, outline planning applications must be screened properly in the first instance to ensure those requiring EIA are identified. Some particular concern has been expressed about planning authorities not requesting sufficient information on outline proposals in order to screen them properly. Planning authorities are therefore reminded of the powers available to them under article 4(3) of the Town and Country Planning (General Development Procedure) (Scotland) Order 1992 (the 1992 Order) in relation to additional details on outline planning applications.

The powers under article 4(3) and under article 13 of the 1992 Order for requesting further information on applications are also of importance in obtaining sufficient information to evaluate the environmental information provided. If the environmental information obtained through the EIA process does not reflect the developer's more detailed proposals, then tying the permission to that environmental information, while necessary, may hamstring the development. It is important to have as full an idea of the proposal as possible, so that, if necessary, further information for the Environmental Statement can be requested (regulation 19 of the EIA Regulations) to ensure it fully reflects the proposal. As a result, restricting the development to the parameters set by the environmental information are less likely to thwart the development.

Should you have any questions relating to this legislation please contact Alan Cameron of the Executive's Planning Division by e-mail at Alan.Cameron@scotland.gsi.gov.uk or by telephone on 0131 244 7065.

Yours sincerely

**JOHN GUNSTONE**

# Environmental Impact Assessment: Questions and Answers

## 1.0    Introduction

**1.1**    Although the Directive has now been in force for many years some planning authorities will have had limited experience of it. This note, in the form of answers to frequently asked questions, offers a brief guide to the Directive, the Regulations and planning authority responsibilities. The guide does not offer definitive guidance and is not a substitute for the Regulations or for guidance provided in the official Scottish Executive Development Department Circular 15/99, Environmental Impact Assessment (Scotland) Regulations 1999 (electronic copy at http://www.scotland.gov.uk/library2/doc04/eia-00.htm), nor the advice in Planning Advice Note 58: Environmental Impact Assessment (electronic copy at http://www.scotland.gov.uk/library/pan/pan58-00.htm). You need to be familiar with these documents and refer to them when dealing with applications where EIA is involved. But it may provide a useful aide-memoire to remind you of some of the potential pitfalls in cases involving EIA and offer some advice on how you can avoid them.

**1.2**    The interpretation of the EIA Directive and Regulations have been aided by a series of court decisions. This paper also includes references to the cases and summarises the judgements. These cases have important implications for the way in which planning authorities exercise their responsibilities. As noted in the covering letter these are English Court cases, but given the similarity of the regulations north and south of the border, they are very relevant to the operation of the EIA Regulations in Scotland.

## 2.0    Background

**2.1**    In the UK, environmental issues have long been taken into account during the planning process. However, practice varied throughout the European Community. Member States agreed in 1985 that procedures should be harmonised so that environmental issues were addressed in a more rigorous, scientific and transparent manner. In 1988 the European Directive on the effects of certain public and private projects on the environment came into effect. The Directive, referred to as the EIA Directive, was amended by Council Directive No. 97/11/EC in 1997. The consolidated text of the directive is reproduced at Appendix 1 of the DETR publication *Environmental Impact Assessment: a guide to procedures*. An electronic copy is available at www.planning.dtlr.gov.uk/eia/guide/index.htm. Please note, however, the guide itself needs to be treated with some caution as it is based largely on the English planning system and regulations.

**2.2**    For projects that are subject to approval through the planning system the requirements of the Directive have been transposed into domestic legislation by the Environmental Impact Assessment (Scotland) Regulations 1999 (SSI 1999 No. 1) (the Regulations). A copy of the Regulations is available at www.scotland-legislation.hmso.gov.uk/legislation/scotland/ssi1999/199900.htm.

## 3.0    What do the Regulations require?

**3.1**    For qualifying projects they require a planning authority to consider, first, whether a proposed project is likely to have a significant effect on the environment. If so, the authority must ensure that the applicant carries out an assessment and prepares and submits to the planning authority a report that identifies, describes and assesses the effects that the project is likely to have on the environment. The process is referred to as Environmental Impact Assessment (EIA), the report as the Environmental Statement (ES).

**3.2**   The ES has to address the direct and indirect effects of the development on a number of factors including the population, fauna, flora, soil, air, water, climatic factors, landscape and archaeology. Full details of the information that has to be included is listed in Schedule 4 to the Regulations. The ES must also contain a non-technical summary so that lay persons can understand what is being proposed and its likely effects.

**3.3**   Members of the public, and statutory consultees, must be given the opportunity to comment on the ES. Before any decision to approve the application may be taken, the planning authority must take into account the ES and any representations made about the environmental effects by the public or consultees. They must also state in their decision that they have done so.

## 4.0     Is there a standard format for an ES?

**4.1**   There is no prescribed format, but in the case of **Berkeley v SSETR** (2000), the House of Lords commented that an ES must not be a paper chase. Lord Hoffman said, 'the point about the Environmental Statement contemplated by the Directive is that it constitutes a single and accessible compilation, produced by the applicant at the very start of the application process, of the relevant environmental information and the summary in non-technical language.'

## 5.0     Do the Regulations apply to all applications for planning permission?

**5.1**   There are 2 classes of project. Schedule 1 of the Regulations lists those for which EIA is mandatory. Schedule 2 lists those where the planning authority is required to consider whether the project is likely to have a significant effect on the environment. Where this is the case, EIA must also be carried out. There is no discretion not to require EIA simply because other information about the project is available.

## 6.0     What action does the planning authority have to take?

**6.1**   The authority's roles involve 'screening' to determine whether a project requires EIA; 'scoping' to advise the applicant of the likely, significant effects on the environment that it wants to see addressed in the ES; consultation with statutory consultees, members of the public and others who may have views on the proposal and the ES; and evaluation of the environmental information presented in the ES and any representations made on it prior to making its decision.

## 7.0     Screening

**7.1**   An applicant for planning permission may ask the planning authority for a 'screening opinion' before submitting the application. If it receives such a request, the authority has to issue an opinion within 3 weeks of the date of receipt unless an extension of time is agreed in writing with the person making the request. A copy of the opinion has to be made available for public inspection where the planning register is kept.

**7.2**    Where a planning application is submitted without an ES, and a screening opinion has not previously been issued, the authority must determine whether the application falls within a class of development listed in either Schedule 1 or 2 of the Regulations and, for any that fall within Schedule 2, whether the project will have a significant effect on the environment. The authority will then issue a 'screening opinion' to the applicant and place a copy on the planning register. Again a period of 3 weeks is allowed from the date the application is received unless an extension of time is agreed in writing with the applicant.

## 8.0     Who has to give the screening opinion?

**8.1**    It is the responsibility of the planning authority to ensure that planning applications are 'screened' to establish whether EIA is required. Normally this will be carried out by the officer dealing with the planning application. But the decision is taken on behalf of the planning authority. If the decision is to be made by officers, it is important to ensure that they have delegated authority to do so.

**8.2**    In *R v St Edmundsbury Borough Council, ex parte Walton* (1999), a decision of the planning authority to grant planning permission was overturned because a decision not to require EIA was taken by an officer who had no formal delegation. PAN 58 gives best practice guidance advice in terms of the management of EIA applications.

## 9.0     What factors are taken into consideration when reaching a screening opinion?

**9.1**    Given their scale and nature, Schedule 1 projects should be easily identified and it is expected the applicant would not submit such a proposal without an ES. But if not, it should be a fairly straightforward matter to decide that EIA is required.

**9.2**    For projects within a category of development listed in Schedule 2 a screening opinion has to be made if the project is located in, or partly within, a 'sensitive area' (as defined in regulation 2(1)) exceeds/meets the criteria/thresholds listed in column 2 of the Table at Schedule 2.

**9.3**    Schedule 3 of the Regulations gives some guidance on how to decide whether these projects are likely to have significant environmental effects. Further indicative guidance is provided in Annexe A of the SEDD Circular 15/99 on Environmental Impact Assessment. Decisions need to be taken on a case-by-case basis. Thresholds shown within the indicative guidance in the Circular are not determinative. Individual projects that fall below the indicative thresholds and criteria in the Regulations may require EIA. The important thing is to consider whether the proposed development is likely to have significant environmental effects and to be clear of the reasons for the decision.

**9.4**    Projects outwith 'sensitive areas' that fall below the thresholds and criteria in Column 2 of the Table at Schedule 2 do not generally require EIA and the authority need not adopt a screening opinion. In effect, the Regulations have already provided a negative screening opinion. The exception to this is where the Scottish Ministers have exercised powers under regulation 4(8) to direct that EIA is required even though it does not meet these thresholds and criteria. Such a direction will usually be in response to a request by the planning authority.

## 10.0 Does the screening opinion have to give reasons for the decision?

**10.1** Where EIA is required, the authority must provide a written statement giving full reasons for its decision. There is no similar requirement where the authority decides that EIA is not required. However, it would be prudent for the authority to make and retain for its own use a clear record of the issues considered and the reason for its decision. This would be very useful in the event of any challenge to the planning decision based on EIA grounds.

## 11.0 Can the screening opinion still be issued outside of the 3 week timescale?

**11.1** To avoid unnecessary delays it is important that every attempt should be made to issue screening opinions within the statutory 3 week period. The regulations do, however, allow for the authority and the applicant to agree a longer period. Unless there is such agreement, the authority has no legal authority to request EIA beyond the 3 week period.

**11.2** But, if it had not issued a screening opinion and it considered that EIA was required, the authority could seek to persuade the applicant voluntarily to carry out an assessment and provide an ES, which would be submitted in accordance with the Regulations. It can also request the Scottish Ministers to issue a screening direction to determine whether EIA is required.

## 12.0 Can the authority change its screening opinion?

**12.1** Yes. But this should be done within the statutory period unless there is prior agreement of the applicant to extend the period.

**12.2** It is possible that additional information about the effects of the project not known to the authority when its screening opinion was given will come to light before a decision is taken on the application. If that information indicates that EIA is required the authority must not ignore it simply because it has already issued an opinion that EIA is not required. If the authority itself is unable to change its opinion, it should request a screening direction from the Scottish Ministers (who have a general power to direct whether EIA is required) before any decision is taken on the application.

**12.3** The case of **Fernback and Others v Harrow LBC** (2000) addressed this issue. In this case the court held that a 'negative' screening opinion issued by a local planning authority (LPA) did not determine whether an application for planning permission was 'EIA Development' and a 'positive' one by the LPA was determinative only in the absence of one by the Secretary of State. On the other hand, an opinion by the Secretary of State, either way, is determinative. In Scotland the role of the Secretary of State would of course be taken by the Scottish Ministers.

## 13.0 Scoping

**13.1** Applicants for planning permission may request the planning authority to provide a 'scoping opinion' on the impacts and issues that the EIA should address – i.e. those impacts that are likely to be significant. The statutory process requires discussion between the authority, applicant and statutory bodies and a scoping opinion to be issued within 5 weeks of the request or such longer period as may be agreed.

**13.2** The Regulations require the authority to issue a scoping opinion only in cases where the application has not yet been submitted. But authorities are encouraged to respond favourably to any request from the applicant for a scoping opinion. They may also wish to consider whether they should extend consultations to involve the public and other interested bodies.

## 14.0    Once a scoping opinion is issued can I request further information?

**14.1** A scoping opinion that is agreed by all interested parties at the outset should ensure that the relevant issues and potential impacts are identified and reported in the ES. Provided EIA is properly carried out this should minimise the need to request further information. However, if the planning authority believes that further information is necessary it is able to request it under regulation 19.

**14.2** It is important to stress that the authority must obtain all the information it needs to assess and evaluate the likely significant environmental effects of the proposal before it reaches its decision. It cannot adopt a 'wait and see' approach or impose a condition requesting further work to identify the likely environmental effects after permission has been granted. It must be sure that all of these have been identified and taken into account before granting planning permission.

**14.3** *R v Cornwall County Council ex parte Jill Hardy* (2001) refers to a case in which the applicant carried out EIA and provided an ES. Although it was known that the conditions at the site were those favoured by a protected species, bats, the applicant did not investigate for their presence. The planning authority, advised by English Nature, granted planning permission but imposed a condition requiring the applicant to carry out a survey to establish whether bats were present prior to commencing the development. The court held that this information should have been included in the ES, otherwise the authority could not comply with the Regulations (regulation 3(2)). The planning permission was quashed.

## 15.0    Who has to be consulted, and when?

**15.1** The Regulations require a planning authority to consult with specified statutory consultees prior to issuing any scoping opinion. It must also give statutory consultees and members of the public an opportunity to comment on any ES and its associated planning application and it must take any relevant views expressed by them into account in reaching its decisions.

**15.2** There is no requirement to consult either statutory consultees or the public about screening opinions.

## 16.0    Do special provisions apply in advertising development subject to EIA?

**16.1** Where an ES is submitted, the planning authority has to advertise this in a local newspaper (and the *Edinburgh Gazette*) and specify where the application and ES may be inspected for a period of 28 days during which time representations can be submitted to the planning authority. The applicant is responsible for the payment of fees relating to the advertising of the application in the newspaper. There is also a specific form of Notice for EIA applications. See Schedule 5 of the Regulations. http://www.scotland-legislation.hmso.gov.uk/legislation/scotland/ssi1999/99900107.htm#sch5

## 17.0    Does further information requested under Regulation 19 also have to be advertised?

**17.1** Yes. The authority will have to advertise in the manner set out in regulation 19, using Schedule 6 of the regulations.    http://www.scotland-legislation.hmso.gov.uk/legislation/scotland/ssi1999/99900107.htm#sch6

## 18.0    What if the applicant changes the ES rather than simply provides further information?

**18.1** There is no specific provision dealing with amendments or additions to an ES that has already been submitted. Such information would not be regarded as 'further information' as this is very specifically defined in the Regulations.

**18.2** The safest approach is to treat any addition or amendment as an ES submitted during the course of a planning application and to advise the applicant to advertise the whole of the ES, with the amendment/addition, in compliance with regulation 13. This will ensure compliance with the general intent of the EIA Directive to notify and inform people of the possible environmental effects of a proposed development.

## 19.0    Evaluating the Environmental Statement

**19.1** The planning authority is responsible for evaluating the ES to ensure it addresses all of the relevant environmental issues and that the information is presented accurately, clearly and systematically. The planning authority should be prepared to challenge the findings of the ES if it believes they are not adequately supported by scientific evidence. If it believes that key issues are not fully addressed, or not addressed at all, it must request further information. The authority has to ensure that it has in its possession all relevant environmental information about the likely significant environmental effects of the project before it makes its decision whether to grant planning permission. It is too late to address the issues after planning permission has been granted.

## 20.0    Does this also apply to applications for outline planning permission where some matters may be reserved for later determination?

**20.1** Yes. Where it applies, the Directive requires EIA to be carried out prior to the grant of 'development consent'. Development consent is defined as 'the decision of the Competent Authority or authorities which entitled the developer to proceed with the development'. Under the UK planning system, it is the planning permission that enables the applicant to proceed with the development. Therefore, in the case of outline applications, EIA applications must be properly assessed for possible environmental effects prior to the grant of outline permission. It will not be possible to carry out EIA at the reserved matters stage. The planning permission and the conditions attached to it must be designed to prevent the development from taking a form – and having effects – different from what was considered during EIA.

**20.2** This was confirmed in the case of *R v SSTLR ex parte Diane Barker* (2001).

## 21.0    For outline planning applications, how should EIA be carried out so as to comply with the Directive and Regulations?

21.1 The cases of *R v Rochdale MBC ex parte Tew* (1999) and *R v Rochdale MBC ex parte Milne* (2000) set out the approach that planning authorities need to take when considering EIA in the context of an application for outline planning permission if they are to comply with the Directive and the Regulations.

**21.2** Both cases dealt with a legal challenge to a decision of the authority to grant outline planning permission for a business park. In both cases an ES was provided. In ex parte Tew the Court upheld a challenge to the decision and quashed the planning permission. In ex parte Milne, the Court rejected the challenge and upheld the authority's decision to grant planning permission.

**21.3** In ex parte Tew, the authority authorised a scheme based on an illustrative masterplan showing how the development might be developed, but with all details left to reserved matters. The ES assessed the likely environmental effects of the scheme by reference to the illustrative masterplan. However, there was no requirement for the scheme to be developed in accordance with the masterplan and in fact a very different scheme could have been built, the environmental effects of which would not have been properly assessed. The Court held that description of the scheme was not sufficient to enable the main effects of the scheme to be properly assessed, in breach of Schedule 4 of the Regulations.

**21.4** In ex parte Milne, the ES was more detailed; a Schedule of Development set out the details of the buildings and likely environmental effects, and the masterplan was no longer merely illustrative. Conditions were attached to the permission 'to tie the outline permission for the business park to the documents which comprise the application'. The outline permission was restricted so that the development that could take place would have to be within the parameters of the matters assessed in the ES. Reserved matters would be restricted to matters that had previously been assessed in the ES. Any application for approval of reserved matters that went beyond the parameters of the ES would be unlawful, as the possible environmental effects would not have been assessed prior to approval.

**21.5** The judge emphasised that the Directive and Regulations required the permission to be granted in the full knowledge of the likely significant effects on the environment. This did not mean that developers would have no flexibility in developing a scheme. But such flexibility would have to be properly assessed and taken into account prior to granting outline planning permission.

**21.6** He also commented that the ES need not contain information about every single environmental effect. The Directive refers only to those that are likely and significant. To ensure it complied with the Directive the authority would have to ensure that these were identified and assessed before it could grant planning permission.

**21.7** The Court of Appeal in ex parte Diane Barker (2001) confirmed this approach.

## 22.0    What are the lessons of these cases?

**22.1** You will want to read these judgements carefully, but there are some general points about applications for outline planning permission:

a. An application for a 'bare' outline permission with all matters reserved for later approval is extremely unlikely to comply with the requirement of the Regulations.

b. When granting outline consent, the permission must be 'tied' to the environmental information provided in the ES, and considered and assessed by the authority prior to approval. This can usually be done by conditions although it would also be possible to achieve this by a planning agreement (under section 75 of the Town and Country Planning (Scotland) Act 1997). An example of a condition was referred to in ex parte Milne (2000). 'The development on this site shall be carried out in substantial accordance with the layout included within the Development Framework document submitted as part of the application and shown on (a) drawing entitled "Master Plan with Building Layouts." The reason for this condition was given as 'The layout of the proposed Business Park is the subject of an Environmental Impact Assessment and any material alteration to the layout may have an impact which has not been assessed by that process.' (see paras 28 and 131 of the judgement).

c. Developers are not precluded from having a degree of flexibility in how a scheme may be developed. But each option will need to have been properly assessed and be within the remit of the outline permission.

d. Development carried out pursuant to a reserved matters consent granted for a matter that does not fall within the remit of the outline consent will be unlawful.

## 23.0    What if I fail to comply with the Regulations?

**23.1** It is possible that proceedings will be initiated by an aggrieved party either through the domestic courts or by reference to the European Commission.

## 24.0    Domestic challenges

**24.1** It should be evident from the court cases referred to that failing to comply with the Regulations may make a decision to grant planning permission unlawful and lead to it being quashed by the court. Although the court has the power not to quash planning decisions where there has been procedural impropriety, this discretion is very limited in cases involving EIA because of the duty to comply with EC legislation. It can only be exercised where there has been 'substantial compliance' with the Directive.

**24.2** If the project is one to which the Regulations apply it is essential to comply fully with them. It is not sufficient to argue that EIA was not necessary because all of the information that could have been in the ES was available elsewhere and was taken into account before the decision was taken; or that had an ES been available the decision would have been the same.

**24.3** In **Berkeley v SSETR** (2000), the House of Lords unanimously emphasised the need to comply with the Regulations. It took the view that when considering compliance with the Regulations it was necessary to consider the EIA Directive. The Lords stressed that the importance of the EIA process extended beyond the decision on the application. Its purpose is to provide individual citizens with sufficient information about the possible effects and give them the opportunity to make representations. The court was not entitled to decide after the decision had been made that the requirement of EIA could be dispensed with on the ground that the outcome would have been the same even if these procedures had been followed. In his leading judgement, Lord Hoffman noted that the Directive did not allow Member States to treat 'a disparate collection of documents produced by parties other than the developer and traceable only by a person with a good deal of energy and persistence as satisfying the requirement to make available to the public the information which should have been provided by the developer'.

## 25.0    Complaints to the European Commission

**25.1** Individuals may, and do, complain to the European Commission that planning applications should have been subject to EIA, or that where EIA was undertaken the procedures were not followed correctly or the information in the Environmental Statement was inadequate. This can lead to formal legal proceedings between the Commission and the United Kingdom. This can be lengthy and prolonged and can increase uncertainty for developers and planning authorities.

## 26.0    How can I avoid legal challenge?

**26.1** Nothing can guarantee there will be no legal challenge. But you can minimise the risk of such challenge being successful by ensuring compliance with all of the Regulations. In particular you should ensure that:
- planning applications are properly screened and copies of screening opinions placed on the planning register;
- Environmental Statements contain all of the information required by Schedule 4 of the Regulations;
- all of the likely significant effects that the project will have on the environment have been identified and taken into account prior to a decision to allow the project to go ahead.

The permission that is granted relates only to the project whose environmental effects have been described, assessed and mitigated in the ES. If the ES describes and assesses the effects of burning a single specific type of fuel in a manufacturing process, the consent for the project should be limited to its operation only with the fuel that has been assessed.

Keep a record of your decisions and why you have reached them.